# Digital Design
## and
## Synthesis
### with
# Verilog® HDL

**Eli Sternheim, Ph.D.**
interHDL, Inc.

**Rajvir Singh**
interHDL, Inc.

**Rajeev Madhavan**
Cadence Design Systems, Inc.

**Yatin Trivedi**
YT Associates

**Design Automation Series**

Automata Publishing Company, San Jose, CA 95129

Cover design: Sam Starfas
Interior design: Kate Rushford Murray
Editing: Martha Cover and Ed Haas

UNIX is a registered trademark of AT&T
Verilog is a registered trademark of Open Verilog International.

Copyright ©1993 by *Automata* Publishing Company

Published by *Automata* Publishing Company

This title is an adaptation of *Digital Design with Verilog HDL*

In addition to this title, the following HDL titles and software
are also available from Automata Publishing Company:

   *1. Quick Reference for Verilog HDL , 26 pages, soft cover, $12*
   *2. Digital Design and Synthesis with VHDL*
   *3. PC/DOS version of a Verilog HDL Simulator, $35 with purchase of this title.*

Please see the *order form* on the last page of the book.

All rights reserved. No part of this book may be
reproduced or transmitted, in any form or by any
means, without written permission from the publisher.

*Automata* Publishing Company
1072 S. Saratoga-Sunnyvale Rd.,
San Jose, CA 95129, USA
Phone: 408-255-0705 Fax: 408-749-8823
email: help@apco.com

Printed in the United States of America
10 9 8 7 6 5 4 3 2

ISBN 0-9627488-2-X

## *Contributing Authors*

**Warren Stapleton**
Nexgen Microsystems, Inc.

**Tom Anderson**
Kubota Pacific Computer, Inc.

**Sudhi Balakrishna**
interHDL, Inc.

*To
our families
and friends
who make this,
and all things,
possible.*

# Foreword

For large systems, *gate-level design is dead*. For over 25 years, logic schematics served as the *lingua franca* of logic design. Not any more. Today, hardware complexity has outrun schematics with chips so complex that schematics show only a web of connectivity, not the functionality of the design. **Engineers are therefore moving toward hardware description languages (HDLs).** The most prominent modern hardware description languages are Verilog and VHDL.

This book is the second text by the same authors to directly **address HDL-based high level design.** This text is adapted for writing synthesizable designs using Verilog-HDL. Once again, it is a designer's book, not a tedious text. With a quick and useful introduction to the language, it steps right off into design with the how, not just the what, of an HDL and synthesis. Its authors are real-life designers who saw a better way to teach people HDL-based design.

Here you will see just how you can use an HDL to describe a design, a real design, not just trivial examples. Included are a simple 32-bit pipelined computer, a cache memory design, a UART, and a floppy disk subsystem. **A separate chapter is devoted to logic synthesis and techniques for writing synthesizable designs.**

In this book you can get a good start on writing synthesizable designs in the Verilog-HDL. Even better, it is extremely well laid out, easy to understand, and fun to read—like a good design.

**HDLs, coupled with logic synthesis, are the stuff of the future. Here, in a simple and straightforward presentation, you can see how to use it. This book is one of the best investments a designer or a student can make.**

Good luck!

Ray Weiss
Technical Editor
EDN

# Preface

As electronic designs get larger and more complex, gate-level descriptions become unmanageable and incomprehensible, making it necessary to express designs in more abstract ways. Just as in the 1970s, when high level programming languages replaced assembly languages, in the 1990s, hardware description languages (HDLs) will replace gate-level schematics. Logic synthesis tools will perform the gate-level implementation. Incorporating an HDL and a synthesizer into design methods is no longer an option but a necessity.

At present, there are two dominant hardware description languages, Verilog and VHDL. Both are public domain languages and have become standards for digital design. The Verilog HDL provides a very concise and readable syntax. Many large pieces of hardware have been successfully designed with Verilog.

**A VHDL adaptation of this book is published as a separate title,** *Digital Design and Synthesis with VHDL* **(Oct 1993).**

With increasing usage of Verilog for system and VLSI design, we saw the usefulness of a text which could help designers in writing Verilog models for simulation and synthesis. Many more designers, who are using either some other HDLs or schematic editors, or are new to the field of hardware design, may want to learn about Verilog in particular and top-down design methods in general.

**This book is the outcome of real-world experience with Verilog. Our intent is to show how to functionally describe pieces of hardware in Verilog using a top-down design approach. This design method is illustrated by taking large system examples. Various pieces of design are grouped into architectural description, synthesizable design, and simulation environment.**

All the models presented here have been simulated and verified using the Verilog-XL simulator of Cadence Design Systems, Inc. and the HDL Compiler of Synopsys, Inc.

We expect our readers to have a background in logic design. Some experience in programming in a high-level language, such as C or any other hardware description language, will certainly help. Models in the text are stripped down to retain simplicity for ease of understanding. However, these models can be extended to incorporate the full functionality of the corresponding design.

Although Verilog provides language constructs for designing at lower levels, namely gate and switch levels, such details are not covered in this text. They are fully covered in the Verilog language reference manual obtainable from Open Verilog International.

We hope that our readers will find this text useful in designing large systems and complex VLSI devices with Verilog HDL. This text should help stimulate growth in the trend of designing with hardware description languages as opposed to schematics. The concept of top-down design illustrated here for Verilog should be applicable to any other language such as VHDL.

## Organization of the Book

The book is organized to present material in a progressive manner. It begins with an introduction to the Verilog HDL and ends with a complete example of modeling and testing a large subsystem. In between it presents modeling concepts of various pieces of hardware: instruction set model of a computer, pipelined CPU control, datapath elements, RAM, CAM, caches, clocks, asynchronous input/output, and a disk controller. Whereas, the instruction set and the disk controller models are written for simulation, all other models are written for synthesis using register transfer level (RTL) subset of Verilog.

Chapter 2, although a good summary of the language, concentrates mainly on the behavior modeling constructs of Verilog. An example is given to illustrate the top-down design method by mixing behavioral descriptions of low-level blocks with structural descriptions of high-level topology.

# Preface

Chapter 3 introduces a 32-bit small instruction set computer, SISC, and describes how to model a VLSI processor at the architecture and register transfer levels. It explains the problem of register interlocks and provides a model for a solution to the problem. Concepts of pipelined control are explained and the corresponding Verilog model is presented for a three-stage pipeline. This model is not synthesizable.

Chapter 4 goes one step further and shows how to model various building blocks of the SISC and its CPU. It takes the architectural description of Chapter 3 and comes up with a top-level structural description of the SISC system. Subsequent sections present functional models for the datapath, control unit, memories, and clocks. All these models are synthesizable.

Chapter 5 discusses cache memories and presents a model of a cache subsystem. Various architectural alternatives for improving the performance of the model are described. Cache memory models are synthesizable.

Chapter 6 discusses modeling an asynchronous input/output device. A dual UART is used as an example. The chapter begins with a structural model of the dual UART and goes on to develop a functional model of the single UART module. A model is presented to provide a scheme for testing the dual UART. Models for single UART are synthesizable.

Chapter 7 discusses a top-down design methodology incorporating logic synthesis. It provides a synthesizable model of the AMD2910 microcontroller. This chapter discusses in details issues of using Verilog for synthesis. Coding style for synthesis and recommendations for optimum synthesis modeling are illustrated.

Chapter 8 presents a complete example of modeling a large device. It takes a floppy disk subsystem and builds models for its controller and disk drive units. A test module is included to test the subsystem. This chapter is a good illustration for the top-down design method and is applicable to any large or small system. Models for the disk subsystem are written for simulation only to provide a test environment and do not represent real design for synthesis.

Chapter 9 provides some useful tips and techniques for modeling and debugging. It shows how to model bidirectional ports, bus transactions in pipelined environments, large memories, loading of interleaved memories, etc.

We have included **full listing of models for all the examples** at the end of each chapter. A condensed language reference manual is provided in Appendix A.

## Acknowledgement

This book has greatly benefited by help from **Greg Ward** and **Mohammed Tatar** of BNR, Canada, **Randy Jewel** and **Srinivas Raghavendra** of Synopsys, Inc., **Jeff Lewis** of Redwood Design Automation, **Shrenik Mehta, Bill Fisher, Dan Curran,** and **Dave Stiles** of Nexgen Microsystems, Inc.

We want to thank the following people who critiqued various aspects of the book during its preparation: **Dr. Suhas Patil** of Cirrus Logic, **Avatar Saini** and **Jay Sethuram** of Intel Corporation, **Maqsood Mannan** of National Semiconductor, **Chetan Saiya** of Tandem Computers, **Waqar Shah** of Advanced Micro Devices, **Spencer Greene** of Nimbus Technology, and **Oana Valcea** of Vertex Semiconductor.

We sincerely wish to thank **Professor Karem Sakallah** of University of Michigan, **Professor Michael D. Ciletti** of University of Colorado at Colorado Springs, **Professor Daniel Gajski** of University of California, Irvine, and Professor **Daniel G. Saab** of University of Illinois at Urabana-Champaign for thoroughly reviewing various parts of the manuscript and providing us with their invaluable feedback.

We sincerely acknowledge the invaluable help we received from the following people of Cadence Design Systems: **Martha Cover** and **Ed Haas** for technical editing, **Sam Starfas** for the cover artwork, and **Amy Witherow** for help in making camera-ready copy of the manuscript. We want to thank **Kate Murray** for the interior design of the book and for final completion of the manuscript in FrameMaker on MacIntosh.

Eli Sternheim<br>
Rajvir Singh<br>
Rajeev Madhavan<br>
Yatin Trivedi<br>
<br>
May 21, 1993

# TABLE OF CONTENTS

| | |
|---|---|
| **Foreword** | vii |
| **Preface** | ix |
|    Organization of the Book | x |
|    Acknowledgement | xii |
| **Why Hardware Description Languages?** | **23** |
|    Evolutionary Trends in Design Methods | 23 |
|    Designing with HDLs | 24 |
|    Designing with Verilog HDL | 25 |
| **Anatomy of the Verilog HDL** | **27** |
|    The Concept of a Module | 27 |
|       Testing the Module | 30 |
|    Data Types | 37 |
|       Physical Data Types | 38 |
|       Abstract Data Types | 38 |

| | |
|---|---|
| The Assignment Statement | 45 |
| The Concept of Time and Events | 50 |
| Time and Event Control | 52 |
| The Concept of Parallelism | 55 |
| The fork-join Pair | 56 |
| The disable Statement | 56 |
| Functions and Tasks | 58 |
| The Behavioral Description | 59 |
| The Structural Description | 61 |
| Mixed Mode Representation | 62 |
| Summary | 63 |
| Exercises | 64 |

## Modeling a Pipelined Processor — 65

| | |
|---|---|
| The SISC Processor Example | 65 |
| Instruction Set Model | 66 |
| Declarations | 68 |
| The Main Process | 71 |
| System Initialization | 72 |
| Functions and Tasks | 73 |
| A Test Program | 75 |
| Running the Model | 75 |
| Debugging | 76 |
| Modeling Pipeline Control | 77 |
| What Is a Pipeline? | 77 |

|    |    |
|---|---|
| Functional Partitioning | 78 |
| The Fetch Unit | 80 |
| The Execution Unit | 81 |
| The Write Unit | 82 |
| Phase-2 Control Operations | 84 |
| The Interlock Problem | 85 |
| Test Vector Generation | 88 |
| Summary | 89 |
| Exercises | 90 |
| **Modeling System Blocks** | **103** |
| Structural Model | 104 |
| Datapath | 107 |
| Incrementer | 108 |
| Adder | 109 |
| Barrel Shifter | 110 |
| Multiplier | 112 |
| Setting Condition Codes | 112 |
| Memories | 113 |
| Random-Access Memory | 114 |
| Content-Addressable Memory | 114 |
| Register File | 115 |
| Clock Generator | 117 |
| Single-Phase Clock | 118 |
| Two-Phase Clock | 118 |
| Clock Driver | 124 |

| | |
|---|---|
| Control Unit | 124 |
| Summary | 125 |
| **Modeling Cache Memories** | **127** |
|     Interfaces | 128 |
|         Processor Interface | 128 |
|         System Bus Interface | 129 |
|     Cache Architecture | 130 |
|     Modeling Component Blocks | 132 |
|         Top Level Module | 133 |
|         Tag RAM Module | 137 |
|         Valid RAM Model | 138 |
|         Data RAM Model | 139 |
|         Tag Comparator | 139 |
|         Data Multiplexors | 140 |
|         Controller Model | 140 |
|         Read Hit | 142 |
|         Write Hit | 142 |
|         Read Miss | 143 |
|         Write Miss | 143 |
|         Wait State Counter | 147 |
|     Testing | 148 |
|     Performance Improvements | 148 |
|         Two-Way Set Associative Cache | 148 |
|         Write Buffering | 150 |
|         Larger Line Size | 150 |

| | |
|---|---|
| Write-Back Policy | 151 |
| Summary | 151 |
| Exercises | 152 |

## Modeling Asynchronous I/O: UART — 163

| | |
|---|---|
| Functional Description of the UART | 163 |
| Functional Model of the Single UART | 168 |
|     Reset Operation | 168 |
|     Clock Generator | 168 |
|     Read Operation | 170 |
|     Write Operation | 170 |
|     Transmit Operation | 171 |
|     Receive Operation | 171 |
| Testing the Dual UART Chip | 173 |
| Implementation of the Single UART | 173 |
| Summary | 179 |

## Verilog HDL for Synthesis — 191

| | |
|---|---|
| Introduction | 191 |
|     What is Synthesis? | 191 |
|     HDL Synthesis | 193 |
|     Synthesis Benefits | 195 |
|     Practical Considerations | 197 |
| Synthesis Design Method | 197 |
|     Design at the Register Transfer Level | 199 |
|     Functional Verification | 199 |

| | |
|---|---|
| Gate Level Implementation | 200 |
| Logic and Timing Verification | 200 |
| Physical Implementation | 201 |
| Design Style for Verilog Synthesis | 202 |
|     States and Event Lists | 202 |
|     Arithmetic and Relational Operators | 204 |
|     Flip-flops | 204 |
|     Delays | 205 |
|     Event Control | 207 |
|     Unknown Values and High Impedances | 208 |
| Traffic Light Controller | 211 |
|     Counter Functional Model | 212 |
|     Finite State Machine (FSM) Functional Model | 213 |
|     FSM Initialization | 213 |
|     FSM Implementation | 213 |
| AMD2910 Microcontroller | 217 |
|     Partitioning the Microcontroller | 217 |
|     Register Model | 219 |
|     Stack Model | 219 |
|     Incrementor Model | 222 |
|     Multiplexor Model | 223 |
|     PLA Model | 224 |
|     Tri-state Out Model | 226 |
|     AMD2910 Structural Model | 228 |
|     Synthesizing AMD2910 | 229 |
| Design Implementation and Management | 230 |

| | |
|---|---|
| Verilog Libraries For Synthesis | 230 |
| Design Planning | 232 |
| Design Partitioning | 234 |
|     Ease of Understanding | 235 |
|     Physical Design Consideration | 235 |
|     Project Management Issues | 235 |
| | 236 |
|     Design Structure | 237 |
|     Module Size Considerations | 239 |
| Simulation and Verification | 241 |
| Summary | 242 |
| Exercises | 243 |
| Source Listing | 248 |
|     Primitive Implementation for 4-bit to 1-bit Multiplexor | 250 |
|     Inefficient 4-bit to 16-bit decoder | 250 |
|     Efficient 4-bit 16-bit decoder | 251 |
|     Timer Functional Model | 252 |
|         Counter Model | 252 |
|         State Model | 253 |
|     AMD2910 Models | 255 |
|         Register Model | 255 |
|         Stack Model | 255 |
|         Incrementor Model | 257 |
|         11-bit Multiplexor | 257 |
|         PLA Model | 258 |

| | |
|---|---|
| Tristate Models | 261 |
| High Impedance Tristate Model | 262 |
| AMD2910 Structural Model | 262 |

## Modeling a Floppy Disk Subsystem — 265

| | |
|---|---|
| Functional Description | 265 |
| Operation of the Disk Subsystem | 267 |
| The Timing Checker | 268 |
| The Floppy Disk Controller | 269 |
| Programmed I/O and DMA transactions | 271 |
| Processing the Controller Commands | 274 |
| The Floppy Disk Drive | 278 |
| Testing the Subsystem | 283 |
| Summary | 285 |
| Source Listing | 286 |

## Useful Modeling and Debugging Techniques — 309

| | |
|---|---|
| Bidirectional Ports | 309 |
| Bus Transactions in a Pipeline Architecture | 312 |
| Combinational Blocks with Unknown Inputs | 316 |
| Large Memory as a Table Lookup | 318 |
| Loading Interleaved Memory | 322 |
| Verification of Setup and Hold Constraints | 325 |
| Effects of Verilog Execution Order and Scheduling | 326 |
| Generation of Test Vectors for Complex Modules | 329 |

| | | |
|---|---|---|
| Verification of the Test Vectors | | 331 |
| Summary | | 333 |

## Condensed Language Reference Manual — 335

| | | |
|---|---|---|
| Syntax Conventions | | 335 |
| Lexical Constructs | | 336 |
| Syntax and Semantics | | 338 |
| User Defined Primitive | | 339 |
| Operation | | 361 |

## Verilog Formal Syntax Definition — 363

# CHAPTER 1

# Why Hardware Description Languages?

### Evolutionary Trends in Design Methods

The use of hardware description languages (HDLs) for logic design has greatly expanded in the last few years. Engineering managers no longer face the dilemma of whether to design with an HDL or not.

Instead, their concern is for selecting a language and incorporating it into their design environment. Designers now prefer to express their design in a functional or behavioral form, deferring the details of implementation to a later stage in the design. An abstract representation helps designers explore architectural alternatives and detect design bottlenecks before detailed design begins.

There are several reasons for the wide acceptability of HDLs over the traditional—and slowly disappearing—way of designing with schematics.

As a result of vastly improved technologies, both chip density and design complexity are steadily increasing. Densities of more than a million transistors on a chip can be attained, and to make such complexity comprehensible to the human mind, it is necessary to express the functionality in a high level language that hides the details of implementation. This is also why high level programming languages replaced assembly languages in large system programs.

The electronics field is becoming more and more competitive. New entrants to the market are generating tremendous pressure to increase efficiency of logic design, to reduce design cost, and most important, to reduce time to market. Extensive simulation can detect design errors before the design is manufactured, thus reducing the number of design iterations. An efficient HDL and the host simulation system have become invaluable in minimizing the number of design errors and have made it possible to have functional chips in the first silicon.

The trend toward larger and more complex designs will continue. In the 1990s we will see designs approaching the million-gate count. Engineers will inevitably design with an HDL and leave the implementation to logic synthesis tools.

## Designing with HDLs

Expressing designs in a hardware description language (HDL) can provide several benefits. An HDL description can be used as a specification of the design. The advantage of using a formal language, such as Verilog® HDL, to specify a design is that the specification is complete and unambiguous. Formal language specification is "soft" compared to the "hard" form of schematics. An HDL representation allows for easy text processing on any word processor or design-specific tools, whereas binary schematic databases usually require a graphics editor or vendor-specific tools.

A second purpose of using an HDL is simulation. Simulating the design can uncover errors that would otherwise be detected only when the hardware is built. Simulation can be performed at several levels. At the functional level, the system is described using high level constructs. At the logic level, the system is described hierarchically where at the bottom of the hierarchy are the basic building blocks. This level can include timing delay information, allowing for timing analysis, whereby setup/hold time can be checked and verified.

A third purpose for using HDL is logic synthesis. There are synthesis tools which can take an HDL description of a design and generate a gate-level implementation with library components. These tools optimize the design with respect to speed, circuit size, or some other cost function. Existing synthesis tools have some limitations; for example, they use only a subset of the language, and the synthesized circuit may not be as efficient as if it were implemented by an expert designer. Still, synthesizing even a portion of the design can save both time and money, and allows for easy prototyping and initial speed/area estimates.

Finally, an HDL is the best way to document a design. A well commented HDL description can give better and more concise documentation than a set of schematics that show gate level details.

## Designing with Verilog HDL

Verilog HDL is simple and elegant. It provides constructs to describe hardware elements in a succinct and readable form. A comparable description, for example in VHDL, can be twice as long as a Verilog description.

In Verilog, a designer needs to learn only one language for all aspects of logic design. Simulation of a design, at least, requires functional models, hierarchical structures, test vectors, and man/machine interaction. In Verilog, all of these are achieved by one language. Almost every statement that can be written in procedural code can also be issued in an interactive session from the terminal.

Verilog is not only concise and uniform, but also is easy to learn. It is very similar to the C programming language. Since C is one of the most widely used programming languages, most designers should be familiar with it and may, therefore, find it easy to learn Verilog.

# CHAPTER 2

# Anatomy of the Verilog HDL

In this chapter we introduce the Verilog hardware description language through a sequence of examples. A more complete specification of the language can be found in the "Language Reference Manual" and in the "Condensed Reference Manual" in Appendix A".

## The Concept of a Module

A module is the basic unit in Verilog. It represents some logical entity that is usually implemented in a piece of hardware. For example, a module can be a simple gate, a 32-bit counter, a memory subsystem, a computer system, or a network of computers. The ports in modules can be single-bit or multiple-bit wide, and each port can be defined as an input, an output or a bidirectional port.

Figure 2.1, Figure 2.2, Figure 2.3 show three possible implementations of a module. The module has the name AND2, and it has three single bit ports, named in1, in2 and out. in1 and in2 are input ports and out is an output port.

```
//structural
module AND2 (in1, in2, out);
      input in1;
      input in2;
      output out;
      wire in1, in2, out;
      and u1 (out, in1, in2);
endmodule
```

Figure 2.1 A module definition example, structural style

```
//data flow
module AND2 (in1, in2, out);
      input in1;
      input in2;
      output out;
      wire in1, in2, out;
      assign out = in1 & in2;
endmodule
```

Figure 2.2 A module definition example, data flow style

```
//behavioral
module AND2 (in1, in2, out);
      input in1;
      input in2;
      output out;
      wire in1, in2;
      reg out;
      always @(in1 or in2)
            out = in1 & in2;
endmodule
```

Figure 2.3 A module definition example, behavioral style

A Verilog model consists of keywords, names, literals, comments and punctuations marks. Verilog is a case sensitive language and all its keywords are lower case. Spaces, tabs and new lines can be used to improve readability. Comments start with /* and end with */.

# Anatomy of the Verilog HDL

All modules start with the keyword `module` followed by the module name, the list of inputs and outputs, and all modules end with the keyword `endmodule`. The module names must be unique. The three modules then declare the type of the inputs and outputs as `wire` except the output of the module in Figure 2.3 which is declared as `reg`. Since all the inputs and outputs in Verilog are defaulted to `wire`, these declarations are not needed (except the one for the `reg` output) and can be omitted. The three modules demonstrate three different modeling techniques, namely structural, data flow and behavioral modeling.

Data flow description is a way of modeling combinational functions. In this paradigm, the function is treated as a directed tree whose leaves are the inputs and whose root is the output. Whenever any of the inputs changes, the output is recalculated and updated. Note again that data flow can only implement combinational functions.

Behavioral modeling is a way of using a high-level language, not unlike software programming languages to describe the hardware. Since the language is very general and powerful, the modules that can be described may have very little to do with the actual hardware that they describe. Indeed, some of the modules may be unrealizable in hardware. This property is both a weakness and a strength of Verilog.

The structural module in Figure 2.1 has one **instance** of an AND gate as well as or, nand, nor, not and others that are called Verilog primitives. They are built into the language and have pre-defined functions. For example, the AND gate instance continuously monitors its inputs, and whenever any of them change, it recalculates its outputs as the AND function of all its inputs.

The data flow module in Figure 2.2 has a single **continuous assignment** statement, indicated by the keyword `assign`. Semantically, a continuous assignment implements a combinational function as follows: all the variables on the right hand side of the = are continuously monitored, and whenever any of them changes - the whole expression is re-evaluated and assigned to the variable on the left hand side of the =. The curious reader may ask: "What happens if two variables change on the right hand side? Is the expression evaluated twice"? Actually, this question is moot. Since the function is combinational (i.e., there is no internal state), the result will be the same whether the expression is evaluated one or more times. It is true that a smart simulator will try to minimize the number of evaluations in order to increase the simulation speed, but as far as the language is concerned, the number of evaluations does not affect the semantics of the continuous assignment construct.

The behavioral module in Figure 2.3 has one **behavioral instance**, as indicated by the keyword `always`. The expression `@(in1 or in2)` instructs the simulator to wait until either `in1` or `in2` has changed, and the assignment:

```
output = in1 & in2;
```

assigns the expression `in1 AND in2` to the variable out. As the `always` keyword implies, once the assignment has been made, the simulator waits for the next change in `in1` or `in2`. In other words, an `always` block behaves like an infinite loop in a conventional programming language.

**Testing the Module**

Writing a Verilog model, like writing a program, is only a small part of the design process. The design has to be tested and verified. Using conventional methods, this means generating test vectors, applying the tests to the inputs of the model, observing the outputs, and comparing them to some expected output patterns. While Verilog supports this method of testing, it encourages the **test fixture** approach. A **test fixture** is a top level module which has no inputs and outputs. The top module instantiates the module(s) that need to be tested plus possibly additional modules that are needed as an environment around the module under test. Regular Verilog (usually behavioral) constructs are used to generate the test vectors and display the results. Figure 2.4 shows a test fixture for our 2 input AND module.

As can be seen, the module `test_and2` has no input and output ports. The module declares internal variables, two `wires` and one `reg`. There is one instance of the module under test, and finally there is a new behavioral construct, `initial`. The `initial` construct is similar to the `always` construct, and both can have behavioral statements inside them, but `initial` block executes only once, whereas an `always` block executes in an infinite loop. Since here we need more than a single statement we enclose all the statements between a `begin` ... `end` pair.

```
module test_and2;
    reg i1, i2;
    wire o;

    AND2 u2 (i1, i2, o);

    initial begin
        i1 = 0; i2 = 0;
        #1 $display("i1 = %b, i2 = %b, o = %b",
           i1, i2, o);
        i1 = 0; i2 = 1;
        #1 $display("i1 = %b, i2 = %b, o = %b",
           i1, i2, o);
        i1 = 1; i2 = 0;
        #1 $display("i1 = %b, i2 = %b, o = %b",
           i1, i2, o);
        i1 = 1; i2 = 1;
        #1 $display("i1 = %b, i2 = %b, o = %b",
           i1, i2, o);
    end
endmodule
```

Figure 2.4 Test Fixture for and2 module

The first line after the begin assigns zero to both i1 and i2. Since i1 and i2 are the inputs to the instance u2, the values will propagate down the hierarchy to the inputs in1 and in2 of the module AND2. AND2 in turn will detect the change on the inputs, it will evaluate the output out, and this value will propagate back up the hierarchy to the wire o in the module top. In the next line of top, the expression: #1 instructs the simulator to wait for one time unit and then proceed. The $display statement causes the simulator to display the variable i1, i2 and o in binary format (%b). The output from simulating top with one of the AND2 modules is shown in Figure 2.5.

```
i1 = 0, i2 = 0, o = 0
i1 = 0, i2 = 1, o = 0
i1 = 1, i2 = 0, o = 0
i1 = 1, i2 = 1, o = 1
```

Figure 2.5 Test results

In Figure 2.6 we will implement a slightly less trivial function, and in the process show a few more features of Verilog. We will implement the function (in1 AND in2) OR (in3 AND in4). This is still a combinational function, but with a delay of 30 time units between any change in the input to the output.

```
module and_or (in1, in2, in3, in4, out);
    input in1, in2, in3, in4;
    output out;
    wire tmp;
    and #10 u1 (tmp, in1, in2), u2 (undec,
            in3, in4);
    or #20 (out, tmp, undec);
endmodule
```

Figure 2.6 Structural and_or module example

The `and_or module` instantiates three Verilog primitive blocks, two and-gates and one or-gate. The `and_or` module in Figure 2.6 shows Verilog's short hand notation for instantiating multiple instances of the same type. The next line shows that a Verilog primitive instance can be anonymous. This is not true for user-defined modules which need to have a name whenever they are instantiated. Although a module instantiation looks very much like a function call in a conventional programming language, there is a basic difference between the two. An instance exists from the beginning of the simulation until the end. There is no way to disable an instance and there is no concept of executing and exiting like there is in conventional functions. Instead, all the instances in a module are continuously executing their internal functionality throughout the simulation. For this reason, the order in which the instances appear in the module is immaterial. For example, we can interchange the last two lines in the `and_or` module without affecting the functionality of the `and_or` module, in the same way that blocks in a schematics can be moved around without affecting the design that it describes.

The outputs of the `and` gates are connected to the inputs of the `or` gates through two wires. One of the wires - `tmp`, was explicitly declared in Figure 2.6. The other wire - `undec`, was declared implicitly, by simply using its name. Implicitly declared wires can be a source of errors, because a misspelled wire name will not be detected during compilation, and instead will be manifested as open circuit during simulation.

The or gate in Figure 2.6 has a delay of 20 time units from input to output, and the and gates have a delay of 10 time units for a total delay of 30 units. The default delay for primitive gates is 0.

The next version of the and_or module (Figure 2.7) uses three continuous assignments to implement the functionality. The first continuous assignment assigns to a wire, tmp, which was declared previously. The second assignment, to tmp1, combines the declaration of the wire with the continuous assignment in a single statement. Note, that there cannot be implicit declarations for wires which are continuously assigned. They have to be explicitly declared using one of the two methods described.

```
module and_or (in1, in2, in3, in4, out);
        input in1, in2, in3, in4;
        output out;
        wire tmp;
        assign #10 tmp = in1 & in2;
        wire #10 tmp1 = in3 & in4;
        assign #20 out = tmp | tmp1;
        // The three assignments could be condensed
        //      into one:
        // assign #30 out = (in1 & in2) | (in3 & in4);
endmodule
```

Figure 2.7 Data flow and_or example

The order of the continuous assignments in the source file has no effect on the functionality (except that wires have to be declared before they are used). Depending on the simulator, the order may affect the efficiency and number of calculations that have to be made, but not the result of the calculations.

For example, suppose that i1 = 0, i2 = 1, i3 = 0 and i4 = 0, and that at some point in time i1 changes to 1 and i2 changes to 0. One simulator may first evaluate tmp to 1 (i1 = 1 and i2 = 1), then evaluate out to 1, then evaluate tmp to 0 and then evaluate out to 0. Another simulator may evaluate tmp to 1, then evaluate tmp to 0 and will not evolute out at all. The first simulator will cause a zero width spike on the output and the second simulator will not. The effects of such a spike are not specified in the language, and models that use such effect are not portable (either between two simulators or between two versions of the same simulator). In other words, if a zero delay spike is the clock input to a flip flop, then the flip-flop may or may not be triggered, depending on the simulator. The behavioral implementation of the module (Figure 2.8) uses an if-then-else construct. Notice again that out has to be declared as `reg` and not `wire`. The differences between regs and wires will be explained in more detail later.

For the moment, note that any variable that is continuously assigned (using the assign keyword) has to be a wire, and any variable that is assigned using = inside an always or initial block has to be of type `reg`.

```
module and_or (in1, in2, in3, in4, out);
    input in1, in2, in3, in4;
    output out;
    reg out;

    always @(in1 or in2 or in3 or in4) begin
        if (in1 & in2)
            out = #30 1;
        else
            out = #30 (in3 & in4);
    end
endmodule
```

Figure 2.8 Behavioral and_or example

```
module test_and_or;
    reg r1, r2, r3, r4;
    wire o;

    and_or u2 (.in2(r2), .in1(r1),.in3(r3),
            .in4(r4), .out(o));

    initial begin : b1
        reg [4:0] i1234;
        for (i1234 = 0; i1234 < 16; i1234 =
            i1234 + 1) begin
            { r1, r2, r3, r4 } = i1234 [3:0];
            #31 $display ("r1r2r3r4 =
                %b%b%b%b, o = %b",
                r1, r2, r3, r4, o);
        end
    end
endmodule
```

Figure 2.9 Test fixture for and_or module

The test fixture module in Figure 2.9 uses several new constructs. The `u2` instance of `and_or` associates the nets to the instance ports by name rather then by position. The form `.i2(r2)` means connect the port `i2` in the instance module (`and_or` in our case) to the expression r2. This notation eliminates the need to know the order of the ports in the instantiated module, and can detect typographical mistakes.

Following the `begin` keyword, we see : b1. This is an optional field which gives a name to the block. A named block may have local variables declared in it. Such local variables shadow any other variable with the same names which are declared outside the block.

In the block b1 we declared a 5-bit wide `reg` variable, i1234. Following this declaration, we have a for loop, very much in the style of the C programming language. Note, however, that the idiomatic i1234++ is not supported, and instead the long notation i1234 = i1234 + 1 has to be used. The variable i1234, although declared as a `reg`, can be used in arithmetic expressions, in which it behaves like an unsigned integer. Inside the loop we use curly braces, { }, which perform concatenation of all the operands between them. A concatenation can appear in any place where a regular variable name can appear, both on the left or the right of the = sign.

The for loop demonstrates an easy way of generating all the combinations of patterns to the inputs i1, i2, i3 and i4. The statement:

```
{ i1, i2, i3, i4 } = i1234[3:0]
```

is equivalent to the four statements:

```
i1 = i1234[3];
i2 = i1234[2];
i3 = i1234[1];
i4 = i1234[0];
```

Since i1234 goes from 0 to 15, its 4 least significant bits go through all their 0/1 combinations, starting from 0000 to 1111. Note that we have to make i1234 5-bits wide, because otherwise the expression: i1234 < 16 will be always TRUE and the loop will never terminate. Figure 2.11 shows the simulation results.

```
               r1r2r3r4 = 0000, o = 0
               r1r2r3r4 = 0001, o = 0
               r1r2r3r4 = 0010, o = 0
               r1r2r3r4 = 0011, o = 1
               r1r2r3r4 = 0100, o = 0
               r1r2r3r4 = 0101, o = 0
               r1r2r3r4 = 0110, o = 0
               r1r2r3r4 = 0111, o = 1
               r1r2r3r4 = 1000, o = 0
               r1r2r3r4 = 1001, o = 0
               r1r2r3r4 = 1010, o = 0
               r1r2r3r4 = 1011, o = 1
               r1r2r3r4 = 1100, o = 1
               r1r2r3r4 = 1101, o = 1
               r1r2r3r4 = 1110, o = 1
               r1r2r3r4 = 1111, o = 1
```

Figure 2.10 and_or simulation result

## Data Types

Like most programming languages, Verilog supports constants which hold fixed data and variables which can be modified during simulation. Constants in Verilog can be decimal, hexadecimal, octal or binary, and have the format:

```
<width>'<radix> <value>
```

where width is an optional decimal integer describing the width of the constant, radix is optional and can be one of b, B, d, D, o, O, h, or H.

Radix B and b indicate a binary constant, radix O and o indicate an octal constant, radix D and d indicate a decimal constant, and radix H and h indicate a hexadecimal constant.

Width is the number of bits in the constant, and is interpreted in decimal. If width is not specified, then it is inferred from the value of the constant, and if radix is not specified then a decimal radix is assumed.

Value is a string of ASCII characters which depending on the radix represent the actual value of the constant. If the radix is b or B then the value characters can be 0, 1, x, X, z or Z. If the radix is o or O the value characters can also be 2, 3, 4, 5, 6, 7. If the radix is h or H, then the value characters can also be 8, 9, a, A, b, B, c, C, d, D, e, E, f, F. For radix d or D the value characters can be any digit, 0 to 9, but they cannot be X or Z. Some examples are:

```
15            (decimal 15)
'h15          (decimal 21, hex 15)
5'b10011      (decimal 19, binary 10011)
12'h01F       (decimal 31, hex 01F)
'b01x         (no decimal value, binary 01x)
```

String constants are written between two double quotes (e.g., "mystring") and are converted to their ASCII equivalent binary format. For example, the string "ab" is equivalent to 16'h5758. Real constants can use scientific notation, e.g. 22.73, 12.8e12.

Verilog supports both abstract data types such as integers, and physical data types which represent actual hardware.

**Physical Data Types**

The physical data type in Verilog are registers and various types of nets. Both registers and nets have a width, specified at declaration, or defaulted to 1. Each bit can have one of four values; 0, 1, X or Z. X represents either a variable that has not been initialized, or a net which has a conflict due to two or more drivers trying to drive it to different values. Z represents high impedance or floating value.

A net can be one of several types; wire, wand, wor, etc. The type of the net determines how its final value is resolved if it is driven by more than one source. The difference between a reg and a net is that a reg remembers the last value assigned to it, whereas a wire needs to be continuously driven. Reg's can be assigned only inside behavioral instances. Nets are driven by either data-flow instances or by outputs of structural blocks, and cannot be assigned in behavioral blocks. Note that a reg does not necessarily imply a latch or a flip-flop in the hardware realization of the model.

Register and nets can be used in arithmetic expressions and are interpreted by the simulator as unsigned integers.

**Abstract Data Types**

Verilog supports the following abstract data types:
- `integer`
- `time`
- `real`
- `event`
- `parameter`.

`integer` is for all practical purposes equivalent to a 32-bit `reg`, with one exception. A `reg` is always treated as a non negative number in arithmetic operations, whereas an `integer` is interpreted as 2's complement signed number.

A `time` variable is similar to integer, except that it is 64-bits wide and is unsigned.

A `real` is a floating point number in the native machine representation. Verilog provides system functions for translating from real to bit-vectors and vice-versa.

An `event` is a special variable that doesn't take any value. They are used to synchronize activities in different parts of the model.

A `parameter` is a named constant. They get set before simulation starts and retain their value for the duration of the simulation. Their type is determined by the value that they are assigned.

**Declarations**

A variable needs to be declared before it can be used. The declaration determines the type and size of the variable. The size of a variable has two components: physical variable have width. `integer`, `real` and `reg` variables can also be declared as arrays. Array is the only data structure available in Verilog.

Here are some examples of declarations of variables:

```
integer i, j;          // two integers
real f, d;             // two real numbers
wire [7:0] bus         // 8-bits wide bus
reg [0:15] word;       // 16-bits wide word
reg arr [0:15];        // array of 16 one-bit reg's
reg [7:0] mem[0:127];  // array of 128 bytes
event trigger,clock_high; // two events
time t_setup,t_hold;   // t1, t2
parameter width=8;
parameter width2=width*2;
wire [width-1:0] ww;
// The following are illegal:
wire w[0:15];          // Wires cannot be in arrays
wire [3:0] a, [7:0] b;
// Only one width specification per declaration
```

Figure 2.11  Basic Operators and Expressions

Variables can be used in expressions which return values. The Verilog language borrows the syntax and semantics of most of its operators from the C programming language. A notable exception is the absence of auto-increment ++ and auto-decrement -- operators. Figure 2.12 gives a summary of Verilog operators, and Figure 2.13 shows the operator precedence.:

```
+ - * /        (arithmetic)
> >= < <=      (relational)
! && ||        (logical)
== !=          (logical equality)
?:             (conditional)
{}             (concatenate)
%              (modulus)
=== !==        (case equality)
~ & |          (bit-wise)
<< >>          (shift)
```

Figure 2.12  Summary of Verilog Operators

```
          *  /  %  Highest precedence
          +  -
          << >>
          <  <= > >=
          == !== === !==
          &
          ^  ^~
          |
          && Lowest precedence
```

Figure 2.13  Operator Precedence

The operands in an expression are variable names (e.g., a), bit-select (e.g., a[i]), part-select (e.g., a[3:0]), or function calls (to be described later). Note that a[i] could mean either the i'th bit of a vector a, or the i'th element in an array a, depending on whether a was declared as:

```
reg [7:0] a;
       or
reg a [7:0];
```

however the expression a[3:2] will only be legal in the first case. Also, note that the index in bit-select can be variable, but for part-select the indices need to be constants or constant expressions. Finally, the order of the bits, low-to-high or high-to-low in part-select need to be the same as it was in the variable declaration.

To access a bit subfield of an array element, the element has to be stored first in a temporary variable

```
reg [15:0] array [0:10];
reg [15:0] temp;
       ...
temp = array[3];
... temp[7:5] ... // Cannot do array[3][7:5]
```

All the operators treat reg's and wires as unsigned (non-negative) integers. This is especially significant for the relational operators. real operands can only be used by the arithmetic and relational operators. Boolean operands are considered TRUE if any of their bits are 1, and FALSE otherwise.

The arithmetic operators +, -, * and / have their usual meaning of addition, subtraction, multiplication and division respectively. If any of the bits of the operands have an X or Z bit, then the result is X. The module operator (%) returns the remainder of dividing the first operand by the second operand.

The relational operators >, >=, <, <= have their usual meaning. The result is one bit value of 1 if the relationship between the operands hold and otherwise 0. If any of the operands has an X or a Z bit, then the result is X.

The logical operators take boolean operands and return one bit values. The logical not operator (!) returns 0 if its operand is TRUE, it returns 1 if its operand is zero, and it returns X otherwise. The logical AND operator (&&) returns 1 if both its operands are TRUE, it returns 0 if any operand is zero, and it returns X otherwise. The logical operator OR operator (||) returns 1 if any of its operands are TRUE, it returns 0 if both operands are zero, and it returns X otherwise.

The logical equality and the case equality operators compare their operands and return one bit. The logical operators return x if the operands contain any X or Z bit. The equivalence operators (== and ===) return 1 if the operands are the same, and return 0 otherwise. The not-equal operators (!= and !==) return 0 if the two operands are equal, and return 1 otherwise. The case-equality operators (=== and !==) can be used to compare operands which have X and Z in them, as shown in Figure 2.14

```
module equequ;
    initial begin
    $display ("'bx == 'bx is %b", 'bx == 'bx);
    $display ("'bx === 'bx is %b", 'bx === 'bx);
    $display ("'bz != 'bx is %b", 'bz != 'bx);
    $display ("'bz !== 'bx is %b", 'bz !== 'bx);
    end
endmodule
```

Figure 2.14 Difference between == and ===

Execution of the module equequ will produce the following results:

```
'bx ==  'bx is x
'bx === 'bx is 1
'bz !=  'bx is x
'bz !== 'bx is 1
```

The conditional operator (:?) takes three operands. If the first operand is TRUE, the operator returns the second operand, otherwise it returns the third operand. The conditional operator can be used to implement a selector:

```
wire w;
assign w = sel ? sig1 : sig2;
```

Nested conditionals can be used to implement multi-way selection:

```
wire [1:0} absval;
assign absval = (a > 0) ? 1:
                (a < 0) ? 2:
                0;
```

The bit-wise operators (~, ^, ^~, &, |)operate on their arguments bit by bit. If the operands are not of the same length, then the shorter one is extended with zeros. The operators return a value which has the width of the wider operand among the two. The operator ~ returns the bit-wise not of its operand. The operator ^ returns the bit-wise exclusive-or of its operands. The operators ~^ and ^~ return the exclusive-nor of their operands. The operator | returns the bit-wise OR of its operand, and the operator & returns the bit-wise AND of its operands.

All the binary bit-wise operators (^, ~^, ^~, &, |)can also be used as unary reductions operators. As unary operators, they operate between the bits of their operand and return one bit results. For example, the expression !var where var is declared as 3 bit variable ([2:0]) is equivalent to var[2]| var[1]| var[0]. A more typical example of using bit-wise operators as reduction operators it the expression ^var. This produces the exclusive-or of all the bits in var or its parity. Other examples are:

```
^word ===1'bx
```

which is TRUE if any of the bits in word is X, and:

```
&word ==0
```

which is TRUE if any of the bits in word is 0;

The left-shift operators (<<) shift the first operand to the left by the number of bits indicated by the second operand. Negative right operand produces right shift. The right-shift operator (>>) shifts the first operand to the right by the number of bits indicated by the second operand. The concatenation operator ( { } ) is one of the most useful and yet overlooked operators in Verilog.

Two or more variables (or constants) can be concatenated by enclosing them in curly braces ({ }), separated by commas:

```
{2'b1x, 4'h7} === 6'b1x0111
```

Concatenations can appear in any expression, or they can appear on the left hand side of an assignment operator. A constant that appears in a concatenation, has to have explicit width (e.g., 1'bz rather than 'bz). The size of the concatenation is the sum of the sizes of its constituents. Concatenations can be used effectively to write concise code. For example:

```
{cout, sum} = in1 + in2 + cin;
```

implements a full adder, and

```
{sreg, out} = {in, reg};
```

implements a shift register.

Replication provides a short form notation for duplicating a constant or a variable. An expression can be replicated by enclosing it in two sets of curly braces and providing the replication factor between the first two opening braces:

```
{3{2'b01}} === 6'b010101
```

Figure 2.15 shows an example of using concatenation to swap two bytes, and using replication to do sign extension of a word.

```
module concat_replicate(swap,signextend);
    input swap,signextend;
    reg[15:0] word;
    reg[31:0] double;
    reg[ 7:0] byte1, byte2;
    initial begin
      byte1 = 5; byte2 = 7;
      if (swap)
          word = {byte2, byte1};
      else
          word = {byte1, byte2};
          if (signextend)
              double = {{16{word[15]}},word};
          else
              double = word;
        end
    endmodule
```

Figure 2.15 Concatenation and replication

Expressions can be combined using the normal precedence rules. Parenthesis can be used to improve the readability and avoid ambiguities.

**Procedural Statements**

While expressions can be used to calculate a values, they cannot be evaluated in isolation, but have to be parts of a statement or other constructs. A simple statement can be an assignment statement or it can be a control flow statement. A compound statement, or a block, consists of a group of statements enclosed within `begin` and `end` or between `fork` and `join`. Statements in a `begin-end` block are executed sequentially, and the whole block is finished when the last statement in it is finished. Statements in a fork-join block are executed concurrently, and the block finishes when the slowest statement in it is finished. Every behavioral instance is composed of a simple or a compound statement.

**The Assignment Statement**

The simple (blocking) assignment statements has one of three forms:
```
lhs-expression = expression;
lhs-expression = #delay expression;
lhs-expression = @event expression;
```

Where the left-hand side (LHS) is a variable name, a bit select, a part select or a concatenation thereof. In the first form, the simulator evaluates the expression on the right-hand side (RHS) and immediately assigns it to the RHS. In the second form, the simulator evaluates the RHS expression, then waits for `delay` and then assigns the value to the LHS. In the third form the simulator waits for the `event` to occur before assigning the RHS to the LHS. In all three cases, the next statement will execute only after the assignment is made.

The non-blocking assignment is similar in syntax to the blocking assignment, except that the = is replaced by <=.
```
lhs-expression <= expression;
lhs-expression <= #delay-expression;
lhs-expression <= @event-expression;
```

Semantically, the difference is that now the RHS is evaluated but is not immediately assigned to the LHS. Instead, it is scheduled to be assigned and in the mean time control continues to flow to the next statement.

Non-blocking assignment is convenient for modeling data flow; For example, when modeling a shift register, the order of statements is important::
```
stage1 = stage2;
stage2 = stage3;
stage3 = stage4;
stage4 = stage5;
```

To eliminate the need to order the statements we can not write:
```
stage1 = #1 stage2;
stage2 = #1 stage3;
stage3 = #1 stage4;
stage4 = #1 stage5;
```

This will only delay the execution, but the order of the statements is still important. Instead we can write:

```
stage1 <= #1 stage2;
stage2 <= #1 stage3;
stage3 <= #1 stage4;
stage4 <= #1 stage5;
```

or even:

```
stage1 <= stage2;
stage2 <= stage3;
stage3 <= stage4;
stage4 <= stage5;
```

**The for Loop**

Figure 2.16 shows the use of a `for_loop`. Execution of the `for_loop` module produces the results in Figure 2.16:

```
module for_loop;
    integer i;
    initial
    for (i = 0; i < 4; i = i + 1) begin
        $display ("i = %0d (%b binary)", i, i);
    end
endmodule
```

Figure 2.16 A `for` statement

```
i = 0 (0 binary)
i = 1 (1 binary)
i = 2 (10 binary)
i = 3 (11 binary)
```

Figure 2.17 Results of `for_loop` execution

**The while Loop**

The effect of the `for_loop` can also be obtained using the while construct as shown in Figure 2.18. Execution of the `while_loop` module produces the same results as produced by the `for_loop` module in Figure 2.16.

```
module while_loop;
      integer i;
      initial begin
            i = 0;
       while (i < 4) begin
       $display ("i = %d (%b binary)", i, i);
       i = i + 1;
       end
   end
endmodule
```

Figure 2.18 A `while` loop statement

**The case Statement**

Figure 2.19 shows the use of the case control structure. .

```
module case_statement;
 integer i;
 initial i = 0;
 always begin
       $display ("i = %0d", i);
       case (i)
             0: i = i + 2;
             1: i = i + 7;
             2: i = i - 1;
             default: $stop;
       endcase
       end
endmodule
```

Figure 2.19 A `case` statement

Execution of the `case` statement module produces the following results:

$$i = 0$$
$$i = 2$$
$$i = 1$$
$$i = 8$$

The selection expression is compared against the `case` expressions, and the match is done on a bit by bit basis (similar to the operator `===`). If none of the cases match, then the `default` case will be executed. If no default case exists, then the execution continues after the case statement. It is a good programming practice to always provide a default clause in a case statement. If the default branch can not occur, then the model may be programmed to print an error message or stop the simulation.

The `casez` and `casex` statements are very similar to the `case` statement with one exception. The `casez` statement handles Z bits as don't care, while the `casex` statement handles both Z and X bits as don't care.

**The repeat Loop**

Verilog has two more control structures which are not very common in other programming languages, the `repeat` and the `forever` constructs. Figure 2.20 describes a `repeat` loop which waits for 5 clock cycles and then stops the simulation.

```
module repeat_loop (clock);
    input clock;
    initial begin
        repeat (5)
        @(posedge clock);
        $stop;
    end
endmodule
```

Figure 2.20 A `repeat` loop

**The forever Loop**

Figure 2.21 shows a `forever` loop, which monitors some condition and displays a message when the condition occurs.

```
module forever_statement(a,b,c);
   input a,b,c;
   initial forever begin
         @(a or b or c)
         if (a + b == c) begin
              $display ("a(%d)+b(%d) =c(%d)"
              ,a,b,c);
              $stop;
              end
       end
endmodule
```

Figure 2.21 A `forever` loop

Even though both the `repeat` and the `forever` statements can be implemented using other control statements, (e.g., the `for` statement), they are very convenient, especially in issuing commands interactively from the keyboard. They have the advantage of not requiring any variables to be declared a priori.

## The Concept of Time and Events

So far, describing behavior in Verilog is more like programming in any structured high level language. The single most important difference between the two is the concept of time and its effect on the execution order of statements in a module. In most programming language, there is a single program counter which indicates the current location of program execution. Since in hardware all the elements operate in parallel, a serial model for execution is not appropriate in an HDL, therefore, Verilog execution is event driven. A global variable designates the simulation time. At each point in time there may be one or more events scheduled to be executed. The event scheduler of the Verilog simulator takes the place of the program counter of a programming language

The Verilog simulator executes all the events scheduled for the current simulation time and removes them from the event list. When no

Anatomy of the Verilog HDL

more events exist for the current simulation time, the simulation time advances to the first element scheduled for the next time. As events are being executed, new events are usually being generated for a future time (or possibly for the current time).

Figure 2.22 depicts the simulation time axis with several events scheduled at different points. Note that the time before the current simulation time can not have any events associated with it, since all the events have been executed and removed.

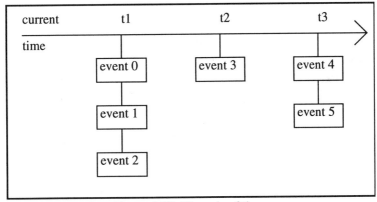

Figure 2.22 The axis of time

The order of event execution within the same simulation time, in general, is not known and one cannot rely on it since the Verilog simulator may try to optimize execution by ordering the events in a particular way. However, Verilog ensures that a linear code which does not have any timing control will execute as a single event without interruption. Also, the order of execution between any two identical

# Digital Design and Synthesis with Verilog HDL

simulation runs will also be identical. Figure 2.23 illustrates this point by using multiple behavioral instances.

```
module event_control;
    reg [4:0] r;
    initial begin
    $display ("First initial block, line 1.");
    $display ("First initial block, line 2.");
    end
    initial
        for (r = 0; r <= 3; r = r + 1)
        $display ("r = %0b", r);
endmodule
```

Figure 2.23 Multiple behavioral instances

Execution of the event_control module produces the following results:

```
First initial block, line 1.
First initial block, line 2.
r = 0
r = 1
r = 10
r = 11
```

Here, all the initial blocks are scheduled to execute at the same simulation time, time 0.

Referring to simulation results of Figure 2.23, one can see that the simulator chose some order of execution of the blocks, in this case, the textural order in the source file. Note that once a block has been scheduled for execution, it continues to execute until completion.

## Time and Event Control

A Verilog process (i.e., an `initial` or `always` block) can reschedule its own execution by using one of the three time control forms:

```
#expression
@event-expression
wait (expression)
```

The `#expression` form suspends the execution of the process for a fixed time period specified by `expression`. The `@event-expression` form, suspends the execution until the specified event occurs. In both cases the scheduler removes the currently executing events from the events list of the current simulation time and puts it on some future events list. The `wait expression` form is a level sensitive event control. If the wait expression is false, execution will be suspended until it becomes true (through the execution of some statement in another process).

These three constructs or any combination thereof can be prepended to any statement. For example, the code:

```
@ (posedge clk) #5 out = in
```

means wait for the positive edge of the clock, then wait for 5 time units and then assign in to out.

Figure 2.24 shows the use of time control structure.

```
module time_control;
    reg [1:0] r;
    initial #70 $stop;
    initial begin : b1 // Note a named block, b1
        #10 r = 1; // wait for 10 time units
        #20 r = 1; // wait for 20 time units
        #30 r = 1; // wait for 30 time units
    end
    initial begin : b2 // Note a named block, b2
        #5 r = 2; // wait for 5 time units
        #20 r = 2; // wait for 20 time units
        #30 r = 2; // wait for 30 time units
    end
    always @r begin
        $display ("r = %0d at time %0d",
                  r, $time);
    end
endmodule
```

Figure 2.24  Example of time control

Execution of the `time_control` module produces the following results:

```
r = 2 at time 5
r = 1 at time 10
r = 2 at time 25
r = 1 at time 30
r = 2 at time 55
r = 1 at time 60
```

Note the first initial statement:

```
initial #70 $stop;
```

which is a common way to limit the simulation run to a fixed time. In this example, even though all the initial blocks started at the same time, some of them were suspended (and rescheduled) at different points on the simulation time. Note the use of named blocks (b1 and b2). A named block can have local variables declared in it, although this example did not exploit this property and used the names as a notational convenience only.

# Anatomy of the Verilog HDL

The `@event-expression` form of control, waits for an event to occur before continuing the execution of the block. An event can be one of several forms:

(a) variable <or variable> ....

(b) posedge one-bit-variable

(c) negedge one-bit-variable

(d) event-variable

In form (a), execution is delayed until any of the variables has changed. In form (b) and (c) execution is delayed until the variable has changed from 0, X, or Z to 1 (if posedge) or from 1, X, or Z to 0 (if negedge). In form (d), execution of the block is suspended until the event is triggered.

An event can be triggered by executing the expression ->event variable. The following example of Figure 2.25 uses event variables to control the order of execution of three initial blocks which execute at the same simulation time.

```
module event_control;
    event e1, e2;
    initial @e1 begin
        $display ("I am in the middle.");
        ->e2;
    end
    initial @e2
    $display ("I am supposed to execute last.");
    initial begin
        $display ("I am the first.");
        ->e1;
    end
endmodule
```

Figure 2.25 Example of event control

Execution of the module `event_control` will produce the following results:

```
I am the first.
I am in the middle.
I am supposed to execute last
```

This form of control ensures the order of execution. Without the event control statement, the Verilog scheduler can choose to schedule the `initial` blocks in arbitrary order.

A special form of the time and event control construct is their use inside an assignment statement. The assignment:

```
current_state = #clock_period next_state;
```

is equivalent to the following two statements:

```
temp = next_state;
#clock_period current_state = temp;
```

and similarly, the assignment:

```
current_state = @(posedge clock) next_-
```

is equivalent to the two statements:

```
temp = next_state;
@(posedge clock) current_state = temp;
```

## The Concept of Parallelism

Verilog has a few other control structures which are not common in other programming languages. One is the `fork_join` construct, and the other is the `disable` statement.

### The fork-join Pair

Figure 2.26 shows an example of the use of the `fork_join` construct.

# Anatomy of the Verilog HDL

```
module fork_join;
      event a, b;
      initial fork
              @a ;
              @b ;
         join
endmodule
```

Figure 2.26  Example of parallel processes

In this example, execution of the `initial` block will be suspended until both events, a and b, are triggered in some order. During the fork, two or more execution threads are activated. When all of them are complete, execution continues at the join. If some of the threads finish before the others, these threads suspend and wait for the rest.

**The disable Statement**

The `disable` statement works like the `break` statement of C. But whereas the `break` statement in a C program only modifies the program counter, the `disable` statement has to remove pending events from the events queues. The `disable` takes as argument a block name and removes the rest of the events associated with this block from the queue. Only named blocks or tasks can be disabled.

Figure 2.27 is a modification of the previous one, but instead of waiting for both the events a and b to occur, this module waits only for one of them, either a or b to occur.

```
module disable_block;
      event a, b; // Block name is needed
      fork : block|
             @a disable block1;
             @b disable block1;
         join
endmodule
```

Figure 2.27  Example of disable

A common use of the `disable` statement is to simulate the `break,` `continue` and `return` statements in C. Figure 2.28 illustrates this.

```
begin : breakloop
  for (i = 0; i < 1000; i = i + 1) begin : continue
      if (a[i] == 0) disable continue;
            // i.e continue
      if (b[i] == a[i]) disable break; // break
      $display ("a[",i,"]=", a[i]);
  end
end
```

Figure 2.28 Simulating break and continue statements

## Functions and Tasks

One of the most powerful modeling techniques in Verilog is the encapsulation of a piece of code in a task or a function. Figure 2.29 shows an example of a task.

```
task tsk;
    input i1, i2;
    output o1, o2;
    $display("Task tsk, i1=%0b, i2=%0b",i1,i2);
    #1 o1 = i1 & i2;
    #1 o2 = i1 | i2;
endtask
```

Figure 2.29 Example of a task

There are a few differences between a task and a function.

A task may have timing control constructs, whereas a function may not. This means that a function executes in zero simulation time and returns immediately (it is essentially combinational). The code that initiated the task has to wait until the task completes execution or until the task is disabled, before continuing the execution. The execution control returns to the statement immediately following the one which initiated the task or the function.

# Anatomy of the Verilog HDL

A task may have both inputs and outputs, whereas a function must have at least one input and does not have any output. A function returns its results by its name.

A task invocation is a statement, whereas a function is invoked when it is referenced in an expression. For example:

```
tsk (out, in1, in2);
```

invokes a task, named `tsk`, and:

```
i = func (a, b, c); // or
assign x = func (y);
```

invokes a function, named `func`.

Figure 2.30 shows an example of a function.

```
function [7:0] func;
        input i1;
        integer i1;
        reg [7:0] rg;
        begin
                rg = 1;
                for (i = 1; i <= i1; i = i + 1)
                rg = rg+1;
                func = rg;
        end
endfuction
```

Figure 2.30  Example of a function

Functions play an important role in logic synthesis. Since functions are combinational, they can be synthesized, and can be used in describing the system. Tasks are a very important tool in organizing the code and making it readable and maintainable. A piece of code that is used more than once should be encapsulated into a task. This helps to localize any change to this part of the code. If a code is expected to be issued interactively from a terminal, it should also be converted to a task in order to save typing. In addition, it is useful to break long procedural blocks into smaller tasks in order to increase the readability of the code.

## The Behavioral Description

Verilog is a top-down design language that supports behavioral description, structural description or a mixed mode description. In the next few sections we will illustrate these aspects of the language with a complete example of a 4-bit adder.

The first example in Figure 2.31 describes the behavior of the 4-bit adder module using high level constructs of Verilog. Note that the sum and zero ports have been declared as registers. This has been done so that a behavioral statement can assign values to them. This description has two behavioral instances, one an `initial` instance and one an `always` instance.

```
module adder4 (in1, in2, sum, zero);
        input [3:0] in1;
        input [3:0] in2;
        output [4:0] sum;
        output zero;
        reg [4:0] sum;
        reg zero;
        initial begin
                sum = 0;
                zero = 1;
        end
        always @(in1 or in2) begin
                sum = in1 + in2;
                if (sum == 0)
                        zero = 1;
                else
                        zero = 0;
        end
endmodule
```

Figure 2.31 Behavioral description of a 4-bit adder

The construct `@(in1 or in2)` causes the simulation to wait until at least one of the inputs -`in1`, `in2` - changes its value. Without this construct the `always` loop will compute forever with the same input values and the simulation time will never advance.

Note that an `always` block can also be programmed as an `initial` block using a `forever` control structure as shown in Figure 2.32.

```
initial begin
    forever
        @(in1 or in2) begin
            sum = in1 + in2;
            if (sum == 0)
                zero = 1;
            else
                zero = 0;
        end
end
```

Figure 2.32 Using an initial forever assignment

Figure 2.33 is a modification of the example from Figure 2.31, and it models adder4 using a continuous assignment.

```
module adder4 (in1, in2, sum, zero);
    input [3:0] in1;
    input [3:0] in2;
    output [4:0] sum;
    reg [4:0] sum;
    output zero;
    assign zero = (sum == 0) ? 1 : 0;
    initial sum = 0;
    always @(in1 or in2)
        sum = in1 + in2;
endmodule
```

Figure 2.33 Using a continuous assignment

Here, `zero` is a `wire` and not a `register`. The conditional operator (?:) replaces the if statement in the previous implementation. The wire declaration of `zero` and its continuous assignment can be combined into a single statement as follows:

```
wire zero = (sum == 0) ? 1 : 0;
```

## The Structural Description

Figure 2.34 implements the 4-bit adder module as a structure of 1-bit full adder sub modules and gates.

```
module adder4 (in1, in2, s, zero);
        input [3:0] in1;
        input [3:0] in2;
        output [4:0] s;
        output zero;
        fulladd u1 (in1[0],in2[0], 0,s[0],c0);
        fulladd u2 (in1[1],in2[1],c0,s[1],c1);
        fulladd u3 (in1[2],in2[2],c1,s[2],c2);
        fulladd u4 (in1[3],in2[3],c2,s[3],s[4]);
        nor u5 (zero,s[0],s[1],s[2],s[3],s[4]);
endmodule
```

Figure 2.34 Structural description of 4-bit adder

In this example, the 4-bit adder is made of four instances of 1-bit adder modules (fulladd) and one instance of nor gate module. This implementation describes the hardware structure and has a one to one correspondence to a schematic. It uses two types of lower level modules: a fulladd and a nor. Even though fulladd is a user-defined module and nor is a Verilog primitive, both are instantiated in the same uniform way.

Note that the structural description implicitly declared three wires: c0, c1 and c2. These wires interconnect the carry bit from one fulladd stage to the input of the next fulladd stage. Verilog permits implicit declaration of single bit wires. If the interconnection is a multiple-bit bus, then an explicit declaration is needed, for example:

```
wire [3:0] databus;
```

Note also that when a module is instantiated, the order of its input/output ports is important. The higher level module adder4 needs to establish a binding between its nets and the corresponding ports of the lower level module `fulladd`. Figure 2.35 shows a simple implementation of the `fulladd` module.

```
module fulladd (in1,in2,carryin,sum,carryout);
    input in1,in2,carryin;
    output sum,carryout;
    assign {carryout,sum} = in1 + in2 + carryin;
endmodule
```

Figure 2.35  Behavior of a 1-bit full adder

## Mixed Mode Representation

The final example in Figure 2.36 describes the `adder4` as a combination of structural and behavioral instances. This model computes the sum output by structural instances of `fulladd` modules whereas it computes the zero output using a behavioral instance.

```
module adder4 (in1, in2, sum, zero);
    input [3:0] in1;
    input [3:0] in2;
    output [4:0] sum;
    output zero;
    reg zero;
    fulladd u1 (in1[0],in2[0], 0,sum[0],c0);
    fulladd u2 (in1[1],in2[1],c0,sum[1],c1);
    fulladd u3 (in1[2],in2[2],c1,sum[2],c2);
    fulladd u4 (in1[3],in2[3],c2,sum[3],
                sum[4]);
    always @(sum )
        if (sum == 0)
            zero = 1;
        else
            zero = 0;
endmodule
```

Figure 2.36  Mixed mode representation

## Summary

To summarize, you can use three modeling styles in Verilog;
- structural

- data-flow
- behavioral

A module can have any combination of structural, data-flow or behavioral instances, and the order of these instances in the module is immaterial. Structural instances are constructed by instantiating lower level modules, which can be either Verilog built-in primitives or user-defined modules. Data-flow instances represent combinational logic and are constructed using the `assign` keyword. Behavioral instance are constructed using either the `initial` keyword or the `always` keyword. The former execute only once, while the latter execute in an infinite loop.

## Exercises

1. A majority gate is a combinational block whose output is the same as the majority of its inputs. Design a three bit majority gate using the three design methods:

    a. Structural, gate level model.

    b. Data flow model (using continuous assignment).

    c. Behavioral model (using the ''always'' block).

2. Write a Verilog model for a test fixture for the majority gate in 1.

3. Run the simulation, exercising the majority gate through all its possible inputs.

CHAPTER

# 3

# Modeling a Pipelined Processor

In this chapter we take the specification of a 32-bit processor and develop a functional model for it through various stages of successive refinement. First we implement an instruction set model, then we describe a register transfer level (RTL) model. In the next chapter we arrive at a structural model that maps the processor to various building blocks. In the process, we explain modeling of such concepts as pipelining, concurrency, instruction execution, functional partitioning, and creation of test vectors.

The emphasis here is on the process of modeling as opposed to describing the architecture of a processor. It is not our intention to explain the detailed functionality of any commercial microprocessor or architecture. Some discussion on processor architecture is presented to explain the concepts and process of modeling.

## The SISC Processor Example

A typical VLSI processor is specified by its architecture and instruction set. Let us define a Small Instruction Set Computer (SISC) that has only ten instructions: load, store, add, multiply, complement, shift, rotate, nop, halt, and branch. We will design a processor that can execute this SISC instruction set.

Before we discuss the implementation of the processor model, we must understand how it executes programs consisting of a mix of these instructions. That is precisely what we expect to learn from an instruction set model.

A block diagram of the SISC system is shown in Figure 3.1, and the instruction set is described in Figure 3.2.

## Instruction Set Model

An instruction set model of a processor describes the effect of executing the instructions and the interactions among them. The

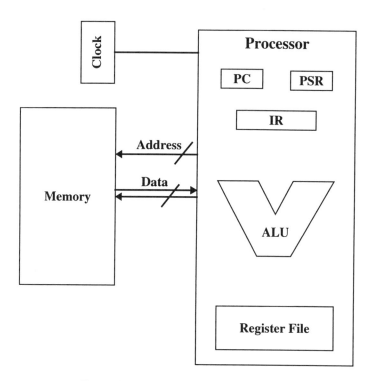

**Figure 3.1** Example SISC block diagram

## Modeling a Pipelined Processor

**Instructions**

| Name | Mnemonic | Opcode | Format | (inst dst,src) |
|---|---|---|---|---|
| NOP | NOP | 0 | NOP | |
| BRANCH | BRA | 1 | BRA | mem, cc |
| LOAD | LD | 2 | LD | reg, mem1 |
| STORE | STR | 3 | STR | mem, src |
| ADD | ADD | 4 | ADD | reg, src |
| MULTIPLY | MUL | 5 | MUL | reg, src |
| COMPLEMENT | CMP | 6 | CMP | reg, src |
| SHIFT | SHF | 7 | SHF | reg, cnt |
| ROTATE | ROT | 8 | ROT | reg, cnt |
| HALT | HLT | 9 | HLT | |

**Condition codes**

| | | |
|---|---|---|
| A | Always | 0 |
| C | Carry | 1 |
| E | Even | 2 |
| P | Parity | 3 |
| Z | Zero | 4 |
| N | Negative | 5 |

**Operand addressing**

```
mem   -  Memory address
mem1  -  Memory address or immediate value
reg   -  Any register index
src   -  Any register index, or immediate value
cc    -  condition code
cnt   -  shift/rotate count,>0=right,<0=left,+/-16
```

**Instruction format**

```
IR[31:28]    Opcode
IR[27:24]    cc
IR[27]       source type 0=reg(mem),1=imm
IR[26]       destination type 0=reg, 1=imm
IR[23:12]    source address
IR[23:12]    shift/rotate count
IR[11:0]     destination address
```

# Digital Design with Verilog HDL

```
Processor status register

PSR[0]   Carry
PSR[1]   Even
PSR[2]   Parity
PSR[3]   Zero
PSR[4]   Negative
```

**Figure 3.2** SISC instruction set

implementation of the corresponding hardware to execute these instructions is not the issue. For example, we can model an add instruction as

```
{carry,sum} = in1 + in2 ;
```

without going into detail of whether the addition is performed using a ripple-carry adder, a carry-look-ahead adder, or some other algorithm. Given two inputs (`in1` and `in2`), an addition is performed to produce the sum and the carry.

Similarly, we are not interested in studying or implementing a memory protocol. We treat memory as a large set of registers directly visible to the processor.

The SISC processor model is a "closed system" module without any input or output ports and has the following form:

```
module system ;
       .... // Module items include
       .... // declarations, tasks, functions,
       .... // initial and always blocks, etc.
endmodule  // system
```

In the following sections we describe each of the module items.

### Declarations

Since we are modeling at a high level, we should think in terms of registers and register bit-fields rather than gates and switches. Some of the registers we need are

```
32-bit register for holding the instructions
12-bit register for addressing the memory
```

# Modeling a Pipelined Processor

```
5-bit register to hold condition code flags
33-bit register to hold results
```

These and other registers and parameters are declared as shown in Figure 3.3.

The memory and the register file are declared as arrays of registers of size `WIDTH`. This declaration allows random access to all locations within the memory and the register file, obviating the need for modeling a communication protocol. Access to a memory or a register file location is now a simple matter of referencing the structure with appropriate location as the array index. For example,

```
RFILE[3] = MEM[20] ;
```

will transfer contents of memory location 21 to register file location 4. Notice that we use 0 as the starting index for both the structures.

The size of the maximum addressable memory was derived from the maximum size of the 12-bit address field. The size of the register file was defined arbitrarily as 16.

```
// Parameter   Declaration

parameter        WIDTH = 32 ;
parameter        CYCLE = 10 ;
parameter        ADDRSIZE = 12 ;
parameter        MAXREGS = 16 ;
parameter        MEMSIZE = (1<<ADDRSIZE);


// Register declarations

reg [WIDTH-1:0] MEM[0:MEMSIZE-1],
                RFILE[0:MAXREGS-1],
                ir,
                src1, src2 ;
reg [WIDTH:0]   result ;
reg [ADDRSIZE-1:0]  pc ;
reg [4:0]       psr ;
reg             dir ;
reg             reset ;

integer         i ;
```

**Figure 3.3** Declarations for instruction set model

The `result` register is defined as 33 bits to hold a carry bit after an arithmetic instruction is executed. The program counter (`pc`), contains the address of an instruction in memory, and therefore has a size of 12. The processor status register, `psr`, holds the five condition code flags: carry, even, parity, zero, and negative.

The parameter statements declare various symbolic constants such as `WIDTH` and `CYCLE`. These symbolic constants allow us to write easily maintainable and customizable models. For example, if the architecture demands 16 kilobytes of memory, we need to change the `ADDRSIZE` parameter from 12 to 14. This will change the `MEMSIZE` parameter from 4k to 16k, and declare the memory array mem with its index range from 0 to a maximum of 16k-1.

The `'define` statements, shown in Figure 3.4, permit us to refer to various fields of the instruction register and the condition codes by symbolic names rather than by numbers. The use of symbolic names for

```
// Define Instruction fields

'define OPCODE     ir[31:28]
'define SRC        ir[23:12]
'define DST        ir[11:0]
'define SRCTYPE    ir[27]
'define DSTTYPE    ir[26]
'define CCODE      ir[27:24]
'define SRCNT      ir[23:12]

// Operand types

'define REGTYPE    0
'define IMMTYPE    1

// Define opcodes for each instruction

'define NOP        4'b0000
'define BRA        4'b0001
'define LD         4'b0010
'define STR        4'b0011
'define ADD        4'b0100
'define MUL        4'b0101
'define CMP        4'b0110
'define SHF        4'b0111
'define ROT        4'b1000
'define HLT        4'b1001
```

**Figure 3.4** Defining symbolic names for bit-fields

various instruction fields in the model description makes the model independent of the arrangement (relative ordering) of these fields in the instruction register. For example, if the order of `SRCTYPE` and `DSTTYPE` fields were exchanged, the entire model will need only two lines changed as follows:

```
'define SRCTYPE ir[26]
'define DSTTYPE ir[27]
```

Figure 3.4 shows definitions of symbolic names for bit fields. Note that the opcode names can be assigned constant values via parameter declarations. In this example, however, there is no need to change the values of the symbolic names once they have been assigned.

**The Main Process**

One approach to understanding the modeling process is to think of it as a three-step manipulation process. First, identify the basic entities or structures to be manipulated. Next, describe how to manipulate these structures. Finally, verify that the model conforms to the specification.

We defined the entities in the previous section: namely, the registers, the register file, and the memory. Now we describe how to manipulate them.

A processor continuously performs a "fetch-execute-write" loop. The Verilog description shown in Figure 3.5 models the main process. The main process, described by the always block labeled `main_process`, is divided into three tasks: `fetch`, `execute`, and `write_result`. The `fetch` task fetches an instruction from memory, the `execute` task executes an instruction, and the

```
always begin : main_process
  if (!reset) begin
     #CYCLE fetch ;
     #CYCLE execute ;
     #CYCLE write_result ;
  end
  else #CYCLE ;
end
```

**Figure 3.5** The main process

write_result task writes the result to the register file. This sequence is repeated for all instructions.

This model assumes that these tasks are sequential and that no pipeline or parallelism is employed. As a result, execution of each instruction takes three cycles to complete. In the next refinement step we will see the implications of a pipeline architecture.

The reset signal (more accurately, the reset register) is checked to see if the processor is being reset. If so, the main process waits for one iteration cycle before checking the reset signal again. The else clause of the if-then-else statement is needed. Without it, the always process becomes a zero delay infinite loop when reset is high.

## System Initialization

The initial state of the simulation model should be equivalent to the state of the real hardware at power-on. To initialize the system, the

```
task apply_reset ;
begin
  reset = 1 ;
  #CYCLE
  reset = 0 ;
  pc = 0 ;
end
endtask

initial begin : prog_load
        $readmemb("sisc.prog",MEM) ;
        $monitor("%d %d %h %h %h",$time,pc,
                 RFILE[0],RFILE[1],RFILE[2]) ;
        apply_reset ;
end
```

**Figure 3.6** Initialization process

test program is loaded, a monitor is set up, and the reset sequence is applied in the initial block as shown in Figure 3.6. The initial block is executed only once at the start of the simulation. It executes concurrently with the always block.

The Verilog system task $readmemb is used to load a test program from an ASCII data file into the memory array MEM.

The reset sequence is applied by invoking the user created `apply_reset` task (Figure 3.6). The reset sequence is very simple in this example; it toggles the `reset` signal and sets the program counter to zero.

The `$monitor` system task provides the debugging information about the simulation time, program counter, and selected registers from the register file.

**Functions and Tasks**

With the declarations, main process, and system initialization in place, the next step is to write the tasks and functions that implement the functionality of the processor. The functions in our description are `getsrc`, `getdst`, and `checkcond`. The tasks are `fetch`, `execute`, `write_results`, `set_condcode`, `clear_condcode`, and `apply_reset`.

The most interesting task is the `execute` task (see Figure 3.7). It uses a case statement to decode the `OPCODE`, and provides appropriate action that corresponds to the execution of each individual instruction.

Note the use of symbolic names to access various fields of the instruction register. If the ordering or the length of various fields were to change, these tasks would not be affected; only the definitions would have to be changed.

The `nop` instruction advances the simulation time by one cycle without any other activity. The halt instruction is implemented using the `$stop` system task. In a real processor, a halt instruction may perform several steps; however, for our purpose of developing this example, it is sufficient to stop the simulation process to recognize that a halt instruction has been executed.

The `default` action provides the mechanism to catch illegal instructions. The `branch` instruction illustrates how a function is called in Verilog.

The load and store instructions implement memory access with a simple protocol. Similarly, the shift and rotate instructions implement a

```
task execute ;
begin
  case ('OPCODE)
    'NOP : ;
    'BRA : begin
             if (checkcond('CCODE)) pc = 'DST ;
           end
    'HLT : begin
             $display("Halt ...") ; $stop ;
           end
    'LD  : begin
             clearcondcode ;
             if ('SRC) RFILE['DST] = 'SRC ;
             else RFILE['DST] = MEM['SRC] ;
             setcondcode({1'b0,RFILE['DST]}) ;
           end
    'STR : begin
             clearcondcode ;
             if ('SRC) MEM['DST] = 'SRC ;
             else MEM['DST] = RFILE['SRC] ;
           end
    'ADD : begin
             clearcondcode ;
             src1 = getsrc(ir) ;
             src2 = getdst(ir) ;
             result = src1 + src2 ;
             setcondcode(result) ;
           end
             . . . // See full listing at the
             . . . // end of the chapter
    default: $display("Invalid Opcode found at ",$time);
  endcase
end
endtask
```

**Figure 3.7** The execute task

fairly complex hardware operation using simple behavioral constructs in Verilog.

The add and multiply instructions at first glance appear very simple. However, that is not the case. The result of adding two numbers can be at most one bit larger than the largest of the two numbers, but the size of the multiplication can be as big as the sum of the sizes of the multiplier and the multiplicand. Assigning the multiplication of two WIDTH size operands to a WIDTH+1 size register (result) saves the least significant WIDTH+1 bits of the result, and only the least significant WIDTH bits are saved in any of the memory or register file locations.

One solution is to use wider registers (2*WIDTH) for computation, and a register pair or two memory locations for storing the result. Another solution is to restrict the size of the operands to no more than half of the width. Yet another approach is to say that the accuracy of the result is the least significant WIDTH+1 bits. In the SISC example we selected 32-bit accuracy of the result.

**A Test Program**

Once the model is written, we must prepare a test program to simulate it. A program to count the number of 1's in a binary pattern is used for this purpose. The program, written in binary, is shown in Figure 3.21. The program is stored in the file sisc.prog and is loaded into the memory array mem using the $readmemb system task. A hexadecimal program can be loaded using $readmemh system task.

**Running the Model**

By executing the test program in our model, we can demonstrate how the instructions are fetched, decoded, and executed, and how the results are stored. This simulation model can now be used to develop other diagnostics, system software, or applications programs for the target hardware. The following command to an operating system prompt will get us started:

```
%verilog sisc.v
```

where sisc.v is the name of the file that contains the Verilog description of the SISC instruction set model.

There are two distinct phases in simulating a model in Verilog simulation environment. The first phase is to eliminate all the compilation and linkage errors. The compilation errors are primarily due to syntax errors and undefined structures. The linkage errors are related to port sizes, module instantiation, missing connections, etc.

The syntactically correct Verilog description of a model can now be simulated to verify its functionality. This phase is usually much more time consuming than the first phase. Use of interactive debugging and some proven techniques can help reduce the time spent here. Some useful debugging techniques are discussed later.

```
//to display register and memory elements
task disprm ;
input rm ;// Display register file or memory
input [ADDRSIZE-1:0] adr1, adr2 ;
begin
  if (rm == 'REGTYPE) begin
     while (adr2 >= adr1) begin
          $display("REGFILE[%d]=%d\n",
                    adr1,RFILE[adr1]) ;
         adr1 = adr1 + 1 ;
     end
  end
  else begin
     while (adr2 >= adr1) begin
        $display("MEM[%d]=%d\n",
                    adr1,MEM[adr1]) ;
        adr1 = adr1 + 1 ;
     end
  end
end
endtask
```

**Figure 3.8** A debugging task to dump states

## Debugging

Although the present model is a correct one, only trivial designs require no debugging. In addition to the interactive debugging environment provided by the simulator, we need some tasks customized for testing a model. For our model, it is important to monitor the changes occurring in the value of the program counter and certain registers in the register file.

An example of a task, disprm, to help debug the model, is shown in Figure 3.8. It is used for displaying the contents of a range of registers or memory locations. It is rather inefficient to type in the body of the task every time there is a need to look at memory locations, say, 20, 21, and 22; issuing separate $display commands is also time consuming when the start and end addresses are more than a few locations apart.

The apply_reset task is useful to "restart" the simulation without having to exit and reenter. It can be made more sophisticated by setting all registers and memory locations to unknowns (x), zero, or a predefined bit pattern. The apply-reset task can also be modified to include the $readmemb system task to load a new program every time a reset is applied to the system.

## Modeling Pipeline Control

So far we have modeled the SISC architecture at the instruction set level. Let us now discuss processor control. We introduce the concept of a pipeline and develop a functional model by refining the SISC model just developed.

### What Is a Pipeline?

Essentially, a pipeline architecture is a way of exploiting inherent parallelism or providing additional resources to create necessary parallelism. A simple pipeline can be built from the inherent concurrency between the fetching of one instruction from memory and the execution of the previously fetched instruction.

In other words, when one instruction is being executed, the next instruction is being fetched. If the next instruction is already available for execution at the end of the current instruction, the processor overlaps the fetch cycle with the execution cycle. This is a typical instruction pipeline.

```
cycle # Without Pipeline    With Pipeline

   1          F1             F1
   2          E1             F2   E1
   3          W1             F3   E2   W1
   4          F2             F4   E3   W2
   5          E2             F5   E4   W3
   6          W2              .   E5   W4
   7          F3              .    .   W5
   8          E3              .    .    .
   9          W3              .    .    .
  10          F4              .    .    .
  11          E4
  12          W4             F12
  13          F5             F13  E12  W11
  14          E5             F14  E13  W12
  15          W5             F15  E14  W13

Without pipeline 5 instructions execute in 15 cycles.
(#instructions)*3

With pipeline 5 instructions execute in 7 cycles.
(#instructions)+2

Legend: F = Fetch,   E = Execute, W = Write
```

**Figure 3.9** Tabular representation of a three-stage pipeline

Figure 3.9 explains how a three-stage pipeline architecture will process the instructions. The stages are fetch(F), execute(E), and write(W). It is assumed that each stage takes one cycle to complete. Therefore, a processor without any pipeline requires three cycles to execute one instruction, whereas a processor with a three-stage pipeline can complete, on average, one instruction every cycle. This concept can be extended to more pipeline stages to improve performance.

Having the next instruction already fetched and available for execution improves performance. However, there are complications and special cases even in this simple concept. For example, a taken branch causes the next instruction to be an instruction different from the instruction at the next memory address. Therefore, an instruction that was already fetched (the prefetched instruction) must be discarded, and the instruction from the branched address should be fetched. Since the processor must discard the prefetched instruction, the execution unit is idle for one cycle—or as many cycles as it takes to fetch an instruction from memory.

In a processor with a larger number of pipeline stages, more complex steps may be required to ensure the proper completion of all the instructions in the pipeline. This phenomenon is generally referred to as "flushing the pipeline." The halt instruction also flushes the pipeline.

Another circumstance that reduces pipeline efficiency arises during the execution of the load and store instructions. Since these instructions are necessary for accessing memory to transfer data to or from the register file inside the processor, the processor cannot fetch the next instruction. The wasted fetch cycle, in turn, causes a wasted execute cycle if the processor has provision for only one prefetch instruction. Multiported memories and register files are used to keep the pipeline full.

**Functional Partitioning**

Now let us improve our earlier SISC design by implementing a three stage pipeline. The three stages of the pipeline are fetch, execute, and write_result. With respect to the instruction currently being executed, the next instruction is being prefetched, and the previous instruction's result is being written in the register file. This is a general scenario that does not hold true for branch, load, and store instructions.

## Modeling a Pipelined Processor

```
always @(posedge clock) begin : phase1_loop
   if (!reset) begin
      fetched = 0 ;
      executed = 0 ;
      if (!queue_full && !mem_access)
         -> do_fetch ;
      if (qsize || mem_access)
         -> do_execute ;
      if (result_ready)
         -> do_write_results ;
   end
end
```

**Figure 3.10** Triggering simultaneous events

We have simplified the SISC processor pipeline by making it synchronous. As shown in the description in Figure 3.10, three simultaneous events—do_fetch, do_execute, and do_write_results—are triggered on the positive edge of the clock (say, phase 1, if it is a two-phase clock). The main_process block of the earlier SISC model is modified to create the phase1_loop block. The negative edge of the clock transfers information between the pipeline stages. Additional declarations required to model the pipeline are shown in Figure 3.11. These registers, wires, and events are referenced in the subsequent explanation of the pipeline model.

```
parameter   QDEPTH = 3 ;// Instr Queue Depth
//   Instr queue, and instr register for write
reg [WIDTH-1:0]    IR_Queue[0:QDEPTH-1], wir ;

// Copy of result, and Execute and fetch pointers
reg [WIDTH:0]   wresult ;
reg [2:0]       eptr, fptr, qsize ;

// Various Controls/flags
reg    mem_access, branch_taken, halt_found ;
reg    result_ready ;
reg    executed, fetched ;
wire   queue_full ;

event  do_fetch, do_execute, do_write_results ;
```

**Figure 3.11** Additional declarations for pipeline modeling

# Digital Design with Verilog HDL

Note that the order of the if statements in the `phase1_loop` is not important because they just trigger the events; the corresponding events and tasks are not necessarily executed in that order. These events trigger the activity of the three functional units.

**The Fetch Unit**

The function of the fetch unit is to transfer an instruction from the memory into the processor and make it available to the execution unit. In order to keep the execution unit continuously busy, we need an instruction queue so that instructions may be fetched a priori. This is achieved by using the `IR_Queue`. We have chosen a queue depth of three to match the number of pipeline stages. A 2-bit register `fptr` refers to the current position in the `IR_Queue` where the next fetched instruction is to be stored. The `qsize` register indicates how many instructions have been prefetched. It is also used to indicate whether `IR_Queue` is empty or full. If the queue is empty, the execution can not proceed, causing wasted execution cycles and a delay in the pipeline; whereas, if `IR_Queue` is full, fetching the next instruction will overwrite an instruction which has not yet been executed.

The `mem_access` flag signals the fetch unit to stall a cycle because a load or a store instruction is in progress. The if statement controlling the `do_fetch` event reflects this. The `fetch` task, shown in Figure 3.12, has been modified to save the fetched instruction in the `IR_queue` instead of loading it directly in the instruction register IR. It

```
task fetch ;
begin
    IR_Queue[fptr] = MEM[pc] ;
    fetched = 1 ;
end
endtask
```

**Figure 3.12** The fetch task

also turns on the `fetched` flag, indicating that a fetch cycle has been completed. The `fetched` flag is used to manipulate queue pointers, discussed later in task set_pointers.

### The Execution Unit

The execution unit, similar to the one described earlier, decodes the current instruction and executes it. It is assumed that all the instructions, with the exception of load and store instructions, can be executed in one cycle. For the same reason, all SISC instructions are single word instructions and all the arithmetic instructions—add, multiply, complement, shift, and rotate—expect the operands to be immediate (part of the instruction word), or contained in the register file.

Further examination of the arithmetic instructions indicates that the ALU requires two input registers, `src1` and `src2`, and one output register, `result`. Since the source as well as the destination of every arithmetic instruction is in the register file, the register file must have three independent ports: two for reading the operands and one for writing the result. Remember that the result of the previous instruction is written in the register file while reading the operands for the current instruction.

The memory is still a scarce resource in our architecture. This effect is modeled in the execution unit as two cycle load and store instructions. The model segment in Figure 3.13 shows how the load instruction is modeled. Upon encountering the load instruction, the execution unit sets the `mem_access` flag to reserve the access to the memory in the next cycle. The fetch unit, in the current cycle, completes reading the next instruction from the memory. The fetch unit is idle in the next cycle, and the execution unit will access the memory. The execution unit essentially executes the load instruction of the previous cycle, because the instruction register is not loaded with the new instruction when `mem_access` flag is set. At the end of the memory transaction, the execute unit resets the `mem_access` flag, allowing the execute unit

```
if (!mem_access) ir = IR_Queue[eptr] ;

'LD : begin
        if (mem_access == 0) // Reserve next
            mem_access = 1 ; // cycle
        else begin    // Mem access
            .......   // in next cycle
        end
      end
```

**Figure 3.13** Memory access for load instruction

```
task flush_queue ;
begin
  // pc is already modified by branch execution
  fptr = 0 ;
  eptr = 0 ;
  qsize = 0 ;
  branch_taken = 0 ;
end
endtask
```

**Figure 3.14** Flushing the pipeline

to get the next instruction from the instruction queue into the instruction register and the fetch unit to fill the instruction queue with new instructions from the memory.

The execution unit stalls a cycle if a branch was taken or the instruction queue was empty. The only two situations that result in an empty instruction queue in the SISC processor are when a branch is taken or when the halt instruction is executed. The `flush_queue` task shown in Figure 3.14 describes the necessary action.

The rest of the model of the execution unit remains unchanged for all the arithmetic instructions as described in the execute task in Figure 3.7. The `set_condcode` and `clear_condcode` tasks and the `checkcond` function are not affected by the pipeline either.

**The Write Unit**

The execution and the write stages of the pipeline communicate via two registers—`wresult` and `wir`. If the result must be written to the register file, the `result_ready` flag is set by the execution unit. The result from the result register is copied into the `wresult` register, and the instruction register (`ir`) is copied into `wir`. The write unit writes the result from the `wresult` register into the destination in the register file as indicated by the `wir` dest field in the next cycle. Since `wir` is a copy of the instruction that computed the result, the dest field correctly reflects the destination in the register file for the previous instruction.

Notice that we simplified the write unit by copying the current instruction from the `ir` register to the `wir` register instead of keeping a pointer to the instruction in the instruction queue. The

```
task copy_results ;
begin
  if (('OPCODE >= 'ADD)&&('OPCODE < 'HLT))begin
      setcondcode(result) ;
      wresult = result ;
      wir = ir ;
      result_ready = 1 ;
  end
end
endtask
```

**Figure 3.15** Copying result and instruction

```
task write_result ;
begin
  if (('WOPCODE>='ADD)&&('WOPCODE<'HLT)) begin
      if ('WDSTTYPE == 'REGTYPE)
           RFILE['WDST] = wresult ;
      else MEM['WDST] = wresult ;
      result_ready = 0 ;
  end
end
endtask
```

**Figure 3.16** Writing result in the register file

`copy_results` task in Figure 3.15 describes how the result from the execution unit is transferred to the write unit, and Figure 3.16 shows how the result is written in the destination register.

Two alternate approaches that use more complex logic can alleviate the need for copying the result from the result register into the result register. One possibility is to modify the result register in the negative clock cycle by the execution unit and copy it into the register file in the positive half of the next cycle by the write unit. Another possibility is to remove the write stage of the pipeline completely and write the result from the ALU directly in the register file. Appropriate simulation analysis can be used to arrive at and justify such decisions.

## Phase-2 Control Operations

Overall, the SISC processor is a synchronous design. The functional blocks, described above, correspond to the three stages of the pipeline. They operate in the positive half of the clock cycle. In the negative half of the clock cycle:

- the program counter (pc) is updated based on whether the branch was taken or not.

- condition codes are set from the newly computed result.

```
task set_pointers ;    // Manage queue pointers
begin
   case ({fetched,executed})
     2'b00 : ;          // idle fetch cycle
     2'b01 : begin      // No fetch
              qsize = qsize - 1 ;
              eptr = (eptr + 1)%QDEPTH ;
            end
     2'b10 : begin      // No execute
              qsize = qsize + 1 ;
              fptr = (fptr + 1) % QDEPTH ;
            end
     2'b11 : begin      // Fetch and execute
              eptr = (eptr + 1)%QDEPTH ;
              fptr = (fptr + 1) % QDEPTH ;
            end
   endcase
end
endtask

always @(negedge clock) begin : phase2_loop
  if (!reset) begin
    if (!mem_access && !branch_taken)
       copy_results ;
    if (branch_taken) pc = 'DST ;
    else if (!mem_access) pc = pc+1; ...
    if (branch_taken || halt_found)
       flush_queue ;
    else set_pointers ;
    if (halt_found) begin
       $stop ;
       halt_found = 0 ;
    end
  end
end
```

**Figure 3.17** Phase-2 control operations

- the instruction register (`ir`) and the `result` register are copied to their shadow registers (the `wir` and the `wresult` registers respectively).
- the fetch and the execute reference pointers (`eptr` and `fptr`) to the instruction queue are updated.

These actions are shown in Figure 3.17 by the `set_pointers` task and the `phase2_loop` process.

The instruction queue requires flushing, which is postponed until the negative half of the cycle because the instructions in the pipeline must be allowed to complete. For the same reason, the simulation of the effect of the halt instruction is not stopped until all the activities in the negative half of the cycle are done. This allows the write unit to store the result of the previous instruction in the positive half of the next cycle when the execution unit sets up the 1-bit `halt_found` flag.

**The Interlock Problem**

There is a problem in the pipeline architecture model that is not readily apparent. The problem, usually referred to as "register interlock,"

```
          Program Segment 1

I1:       ADD      R1, R2    // R1 = R1 + R2
I2:       CMP      R3, R2    // R3 = ~R2

          Program Segment 2

I3:       ADD      R1, R2    // R1 = R1 + R2
I4:       CMP      R3, R1    // R3 = ~R1
```

**Figure 3.18** The interlock problem

appears when an instruction that modifies the contents of a register in the register file is followed too closely by another instruction that attempts to read the same register. The manifestation of this behavior can be more complex and difficult to recognize in an architecture with more pipelined stages.

The register interlock situation is best explained by an example. Consider the two program segments shown in Figure 3.18. In program segment 1, there is no contention of the resource, the R2 register in the register file, provided the register file is multiported.

The scenario changes in program segment 2. The first instruction (I3) reads two registers (R1 and R2) from the register file and writes the result of the addition in register R1. The second instruction (I4) reads the value of register R1 and stores its complement in R3. Due to the concurrency between the execution unit and the write unit, the execution unit may read an incorrect value of R1 while executing I4. (The result of I3 would be in the process of being written to the register R1.)

If the value of R1 is read before the write is complete, I4 will operate on the previous value of register R1. In the real hardware, this could be a hazard or a race condition, depending on the implementation of the read and the write operation for the multiported register file. Two alternate solutions to the problem are discussed below.

```
I3:     ADD     R1, R2  // R1 = R1 + R2
IX:     NOP             // Avoid interlock
I4:     CMP     R3, R1  // R3 = ~R1
```

**Figure 3.19** Modified program segment 2

The simplest solution is to insert the nop instruction between the pair of instructions that cause register interlock. The advantage is that no modification of the architecture and the design is required. The disadvantage, however, is the execution of additional instructions, wasting one cycle for every register interlock and reducing the throughput. It would also require modification of the software, in particular the SISC compiler and the optimizer. The modified program segment 2 is shown in Figure 3.19.

An alternate approach is to implement additional logic in the design such that the register interlock conditions are recognized. Upon recognizing the situation, the hardware can copy the contents of the result register from the previous instruction to one of the operands (src1 or src2) for the current instruction. Simultaneously, reading the operand from the interlocking register must be prevented to avoid overwriting with an incorrect value. This is known as the "bypassing" technique. The

## Modeling a Pipelined Processor

```
reg      bypass ;

function [31:0] getsrc;
input [31:0] i ;
begin
    if (bypass)  getsrc = result ;
    else if ('SRCTYPE === 'REGTYPE)
             getsrc = RFILE['SRC] ;
    else getsrc = 'SRC ;   // immediate type
end
endfunction

function [31:0] getdst;
input [31:0] i ;
begin
    if (bypass)  getdst = result ;
    else if ('DSTTYPE === 'REGTYPE)
             getdst = RFILE['DST] ;
    else $display("Error : Immediate data
             cannot be destination.") ;
end
endfunction

always @(do_execute) begin : execute_block
    if (!mem_access) begin
        ir = IR_Queue[eptr] ;
        bypass = (('SRC == 'WDST) ||
                  ('DST == 'WDST)) ;
    end
    execute ;
    if (!mem_access) executed = 1 ;
end
```

**Figure 3.20** Modified functions and execute process for bypass

model segment shown in Figure 3.20 implements the bypassing in SISC to remove the interlocking problem.

Similar care should be taken for the interlock created when the destination register of an instruction is used as the source for the store instruction that follows immediately. Simulation analysis should be used in conjunction with the design goals to decide which alternative to implement.

**Test Vector Generation**

Once a model of any system is developed, it is important to test it in order to establish its validity. Some nontrivial problems related to the architecture or the implementation, such as the register interlock, are uncovered during testing. This is equally true with models of a simple NAND gate, a complex microprocessor, a complicated microprocessor-based computer systems and even networks of such systems. Simple techniques are applied when the models are small, easy to comprehend, and exhaustive testing is possible. In the case of complex systems, it is not always feasible or practical to generate exhaustive tests. Rather, a large number of diagnostic programs is used as test vectors to test certain sections of the system.

These diagnostic programs can be written in a high level programming language such as C, and a compiler written for the target system can generate corresponding object code (machine code). If a compiler is not available, a simple assembler for the target instruction set can be used to generate machine code from hand-coded assembly language programs. Such tools ease the cumbersome, time-consuming, and error-prone task of creating test vectors in 1's and 0's. A further disadvantage of hand-coding is often learned through painful experience when the instruction format (order of various instruction fields) changes.

Once the diagnostic program has been translated to a sequence of instructions, it is loaded into the simulation model's memory, just as an executable program is loaded into the system's memory. Applying a reset to the system starts the instruction fetch with the first instruction pointed to by the program counter. The instruction itself is equivalent to applying an external test vector to the pins of the processor. The execution of the instruction inside the processor controls the fetching of the subsequent instructions (test vectors).

The Verilog-XL simulator can load a diagnostic program saved in an ASCII file into a simulated memory by using one of the two system tasks: `$readmemb` or `$readmemh`. The `$readmemb` task is used when the instructions are represented in binary, and `$readmemh` is used when the instructions are represented in hex. The program file may contain comments, underbars (_), and white space to enhance readability.

```
// Program to count number of 1's in a given binary number.
//
0010_1000_0000_0000_0000_0000_0000_0001  //LD R1,#0
0010_0000_0000_0000_1001_0000_0000_0000  //LD R2,NMBR
0001_0010_0000_0000_0000_0000_0000_0100  //STRT:BRA L1
0100_1000_0000_0000_0001_0000_0000_0001  //ADD R1,#1
0111_1000_0000_0000_0001_0000_0000_0000  //L1:SHF R2, #1
0001_0100_0000_0000_0000_0000_0000_0111  //BRA L2, ZERO
0001_0000_0000_0000_0000_0000_0000_0010  //BRA STRT, ALW
0011_0000_0000_0000_0001_0000_0000_1010  //L2:STR RSLT,R2
1001_1111_1111_1111_1111_1111_1111_1111  //HLT
0101_0101_0101_0101_1010_1010_1010_1010  //NMBR:5555aaaa
0000_0000_0000_0000_0000_0000_0000_0000  //RSLT:00000000
```

**Figure 3.21** An assembly language program as test vectors

Typically, the program is loaded in the simulated memory at the beginning of a simulation run. This is done using the $readmemb system task in an initial statement. (See the block labeled prog_load in Figure 3.6). A sample program that computes the number of 1's in a binary number is shown in Figure 3.21. Here, the end of the program is indicated by stopping the simulation to allow examination of registers, memory locations, and other signals and variables. In order to speed up the process of verifying an error-free execution of the program, it is a common practice to check computed results against the expected results. In a diagnostic program consisting of hundreds or thousands of instructions, such comparisons may be scattered throughout many parts of the program.

When the simulation model is very large, leaving and reentering the simulator to load a new diagnostic program can be very time-consuming. If the behavior of a reset is modeled to bring the machine to its initial state, the $readmemb or $readmemh system task can be used interactively to load a new program at the end of the current program.

## Summary

This chapter discussed how to model a VLSI processor in Verilog HDL. The model was refined from its initial architectural description without a pipeline to a high level functional description with three pipeline stages. The register interlock problem was discussed at length to show the usefulness of simulation in solving problems or making design decisions. Concepts of test vector creation were also presented.

The complete Verilog HDL descriptions of both models are provided in Figure 3.22.

## Exercises

These exercises are in the increasing order of complexity.

1. Write a program in RSISC assembly language to implement step-wise multiplication algorithm (add and shift). Define three memory locations MULTIPLIER, MULTIPLICAND, and RESULT at the end of your instructions. Initialize MULTIPLIER and MULTIPLICAND with 19 and 23 respectively. Manually convert the assembly code into binary (1's and 0's) and save it in a file named sisc.prog. Run this program on the rsisc.v model to verify the results.

2. Modify the program in Exercise 1 to compare the result of step-wise multiplication with the result of MUL instruction. Define a memory location, COMPARE, which will get 1 if the two results agree and 0 if they do not.

3. Re-run exercises 1 and 2 on the pipeline model in rsisc_pipe.v.

4. Add logic-group instructions, AND, OR, XOR to the RSISC instruction set. Modify the instruction format as necessary. Make sure the status flag is updated after the execution of these instructions.

    Write a program to demonstrate the use of these instructions.

5. Add Indexed Offset addressing mode to the instruction format. The Indexed Offset address is computed by adding the content of the register indexed by SRC or DST field (as appropriate) to the address in the word following the instruction.

    Make necessary changes in the instruction format and the rsisc.v model. Create a new function or a task to compute the effective address. If possible, combine the other addressing modes into this function.

Using the new addressing mode, you can access a much larger memory space. Write a program to demonstrate "long jump".

6. Assume that the memory is implemented with a 2-way interleaved memory bank. Change the memory model in rsisc.v and rsisc_pipe.v to reflect the new architecture.

7. Repeat Exercise 6 if the memory was separated into data memory and instruction memory.

8. Rewrite models in rsisc.v and rsisc_pipe.v so that they can be synthesized.

9. For real enthusiasts - Write a simple assembler to generate binary code from the assembly instruction code.

10. Pick your favorite microprocessor (e.g. Intel 486, SPARC, R4000, etc.). Write an instruction set model for it. Take an assembly instruction program and run it on your model to verify the functional correctness of the model.

## Processor Model Without Pipeline

```
/* ==================================================
 *
 * Model of SISC without piepline.
 *
 * sisc.v
 *
 * Useful for instruction set simulation.
 * Three main tasks - fetch, execute, write.
 *
 * %W% %G%   --  For version control
 */

module instruction_set_model ;

/*
 * Declarations, functions, and tasks
 `*/

// Declare parameters

parameter CYCLE = 10 ;            // Cycle Time
parameter WIDTH = 32 ;            // Width of datapaths
parameter ADDRSIZE = 12 ;         // Size of address fields
parameter MEMSIZE = (1<<ADDRSIZE);// Size of max memory
parameter MAXREGS = 16 ;          // Maximum registers
parameter SBITS = 5 ;             // Status register bits

// Declare Registers and Memory

reg [WIDTH-1:0] MEM[0:MEMSIZE-1], // Memory
                RFILE[0:MAXREGS-1], // Register File
                ir, // Instruction Register
                src1, src2 ; // Alu operation registers
reg [WIDTH:0]   result ; // ALU result register
reg [SBITS-1:0] psr ; // Processor Status Register
reg [ADDRSIZE-1:0] pc ; // Program counter
reg             dir ; // rotate direction
reg             reset ; // System Reset
integer         i ; // useful for interactive debugging

// General definitions

`define TRUE    1
`define FALSE   0

`define DEBUG_ON     debug = 1
`define DEBUG_OFF    debug = 0

// Define Instruction fields

`define OPCODE  ir[31:28]
`define SRC     ir[23:12]
```

## Modeling a Pipelined Processor

```verilog
`define DST     ir[11:0]
`define SRCTYPE ir[27] //source type,0=reg (mem for LD),1=imm
`define DSTTYPE ir[26] //destination type, 0=reg, 1=imm
`define CCODE   ir[27:24]
`define SRCNT   ir[23:12]//Shift/rotate count -=left, +=right

// Operand Types

`define REGTYPE    0
`define IMMTYPE    1

// Define opcodes for each instruction

`define NOP     4'b0000
`define BRA     4'b0001
`define LD      4'b0010
`define STR     4'b0011
`define ADD     4'b0100
`define MUL     4'b0101
`define CMP     4'b0110
`define SHF     4'b0111
`define ROT     4'b1000
`define HLT     4'b1001

// Define Condition Code fields

`define CARRY   psr[0]
`define EVEN    psr[1]
`define PARITY  psr[2]
`define ZERO    psr[3]
`define NEG     psr[4]

// Define Condition Codes

                      // Condition Code set when ...
`define CCC    1      // Result has carry
`define CCE    2      // Result is even
`define CCP    3      // Result has odd parity
`define CCZ    4      // Result is Zero
`define CCN    5      // Result is Negative
`define CCA    0      // Always

`define RIGHT  0      // Rotate/Shift Right
`define LEFT   1      // Rotate/Shift Left

// Functions for ALU operands and result

function [WIDTH-1:0] getsrc;
input [WIDTH-1:0] in ;
begin
  if (`SRCTYPE === `REGTYPE) begin
      getsrc = RFILE[`SRC] ;
  end
  else begin // immediate type
      getsrc = `SRC ;
  end
```

```verilog
   end
endfunction

function [WIDTH-1:0] getdst;
input [WIDTH-1:0] in ;
begin
  if ('DSTTYPE === 'REGTYPE) begin
      getdst = RFILE['DST] ;
  end
  else begin // immediate type
    $display("Error:Immediate data can't be destination.");
  end
end
endfunction

// Functions/tasks for Condition Codes

function checkcond;      // Returns 1 if condition code is set.
input [4:0] ccode ;
begin
  case (ccode)
    'CCC : checkcond = 'CARRY ;
    'CCE : checkcond = 'EVEN ;
    'CCP : checkcond = 'PARITY ;
    'CCZ : checkcond = 'ZERO ;
    'CCN : checkcond = 'NEG ;
    'CCA : checkcond = 1 ;
  endcase
end
endfunction

task clearcondcode ;     // Reset condition codes in PSR.
begin
  psr = 0 ;
end
endtask

task setcondcode ;// Compute the condition codes and set PSR.
input [WIDTH:0] res ;
begin
  'CARRY = res[WIDTH] ;
  'EVEN  = ~res[0] ;
  'PARITY = ^res ;
  'ZERO  = ~(|res) ;
  'NEG = res[WIDTH-1] ;
end
endtask


// Main Tasks - fetch, execute, write_result

task fetch ;    // Fetch the instruction and increment PC.
begin
  ir = MEM[pc] ;
  pc = pc + 1 ;
end
```

```
endtask

task execute ;   // Decode and execute the instruction.
begin
  case ('OPCODE)
    'NOP : ;
    'BRA :begin
            if (checkcond('CCODE) == 1) pc = 'DST ;
          end
    'LD : begin
            clearcondcode ;
            if ('SRCTYPE) RFILE['DST] = 'SRC ;
            else RFILE['DST] = MEM['SRC] ;
            setcondcode({1'b0,RFILE['DST]}) ;
          end
    'STR :begin
            clearcondcode ;
            if ('SRCTYPE) MEM['DST] = 'SRC ;
            else MEM['DST] = RFILE['SRC] ;
          end
    'ADD :begin
            clearcondcode ;
            src1 = getsrc(ir) ;
            src2 = getdst(ir) ;
            result = src1 + src2 ;
            setcondcode(result) ;
          end
    'MUL :begin
            clearcondcode ;
            src1 = getsrc(ir) ;
            src2 = getdst(ir) ;
            result = src1 * src2 ;
            setcondcode(result) ;
          end
    'CMP :begin
            clearcondcode ;
            src1 = getsrc(ir) ;
            result = ~src1 ;
            setcondcode(result) ;
          end
    'SHF :begin
            clearcondcode ;
            src1 = getsrc(ir) ;
            src2 = getdst(ir) ;
            i = src1[ADDRSIZE-1:0] ;
            result = (i>=0) ? (src2 >> i) : (src2 << -i);
            setcondcode(result) ;
          end
    'ROT :begin
            clearcondcode ;
            src1 = getsrc(ir) ;
            src2 = getdst(ir) ;
            dir = (src1[ADDRSIZE-1]==0) ? 'RIGHT : 'LEFT ;
            i = (src1[ADDRSIZE-1]==0) ?
```

```verilog
                        src1 : -src1[ADDRSIZE-1:0];
                    while (i > 0) begin
                        if (dir == 'RIGHT) begin
                            result = src2 >> 1 ;
                            result[WIDTH-1] = src2[0] ;
                        end
                        else begin
                            result = src2 << 1 ;
                            result[0] = src2[WIDTH-1] ;
                        end
                        i = i - 1 ;
                        src2 = result ;
                    end
                    setcondcode(result) ;
              end
       'HLT :begin
                    $display("Halt ...") ;
                    $stop ;
              end
       default : $display("Error : Illegal Opcode.") ;
    endcase
  end
end
endtask

// Write the result in register file or memory.
task write_result ;
begin
    if (('OPCODE >= 'ADD) && ('OPCODE < 'HLT)) begin
        if ('DSTTYPE == 'REGTYPE) RFILE['DST] = result ;
        else MEM['DST] = result ;
    end
end
endtask


// Debugging help ....

task apply_reset ;
begin
    reset = 1 ;
    #CYCLE
    reset = 0 ;
    pc = 0 ;
end
endtask

task disprm ;
input rm ;
input [ADDRSIZE-1:0] adr1, adr2 ;
begin
    if (rm == 'REGTYPE) begin
        while (adr2 >= adr1) begin
            $display("REGFILE[%d]=%d\n",adr1,RFILE[adr1]) ;
            adr1 = adr1 + 1 ;
        end
    end
```

## Modeling a Pipelined Processor

```
      else begin
         while (adr2 >= adr1) begin
            $display("MEM[%d]=%d\n",adr1,MEM[adr1]) ;
            adr1 = adr1 + 1 ;
         end
      end
   end
endtask

// Initial and always blocks

initial begin : prog_load
   $readmemb("sisc.prog",MEM) ;
   $monitor("%d %d %h %h %h",
            $time,pc,RFILE[0],RFILE[1],RFILE[2]);
   apply_reset ;
end

always begin : main_process
   if (!reset) begin
      #CYCLE fetch ;
      #CYCLE execute ;
      #CYCLE write_result ;
   end
   else #CYCLE ;
end
endmodule
```

## Processor Model Without Pipeline

```verilog
/* ======================================
 *
 * Model of SISC processor with pipeline
 *
 * sisc_pipeline_model.v
 *
 * %W%   %G%   --  For version control
 */

module pipeline_control ;

// Declare parameters

parameter CYCLE = 10 ;              // Cycle Time
parameterHALFCYCLE = (CYCLE/2) ;    // Half Cycle Time
parameter WIDTH = 32 ;              // Width of datapaths
parameter ADDRSIZE = 12 ;           // Size of address fields
parameter MEMSIZE = (1<<ADDRSIZE);  // Size of max memory
parameter MAXREGS = 16 ;            // Maximum registers
parameter SBITS = 5 ;               // Status register bits
parameterQDEPTH = 3 ;               // Instruction Queue Depth

// Declare Registers and Memory

reg [WIDTH-1:0] MEM[0:MEMSIZE-1], // Memory
                RFILE[0:MAXREGS-1], // Register File
                ir, // Instruction Register
                src1, src2 ; // Alu operation registers
reg [WIDTH:0]   result ; // ALU result register
reg [SBITS-1:0] psr ; // Processor Status Register
reg [ADDRSIZE-1:0] pc ; // Program counter
reg             dir ; // rotate direction
reg             reset ; // System Reset
integer         i ; // useful for interactive debugging

// Declare additional registers for pipeline control

reg [WIDTH-1:0] IR_Queue[0:QDEPTH-1],   // Instruction Queue
                wir; // Instruction Register for write stage
reg [2:0]       eptr, fptr, qsize ; // Book keeping pointers
reg             clock ; // System Clock
reg [WIDTH:0] wresult ; // Alu result register for write stage

// Various Flags - control lines

reg             mem_access, branch_taken, halt_found ;
reg             result_ready ;
reg             executed, fetched ;
reg             debug ;

wire            queue_full ;

event           do_fetch, do_execute, do_write_results ;
```

# Modeling a Pipelined Processor

```
// General definitions

'define TRUE    1
'define FALSE   0

'define DEBUG_ON        debug = 1
'define DEBUG_OFF       debug = 0

// Define Instruction fields

'define OPCODE  ir[31:28]
'define SRC     ir[23:12]
'define DST     ir[11:0]
'define SRCTYPE ir[27] //source type,0=reg (mem for LD),1=imm
'define DSTTYPE ir[26] //destination type, 0=reg, 1=imm
'define CCODE   ir[27:24]
'define SRCNT   ir[23:12]//Shift/rotate count -=left, +=right

// Defines for Write instructions

'define WOPCODE wir[31:28]
'define WSRC    wir[23:12]
'define WDST    wir[11:0]

// Operand Types

'define REGTYPE    0
'define IMMTYPE    1

// Define opcodes for each instruction

'define NOP     4'b0000
'define BRA     4'b0001
'define LD      4'b0010
'define STR     4'b0011
'define ADD     4'b0100
'define MUL     4'b0101
'define CMP     4'b0110
'define SHF     4'b0111
'define ROT     4'b1000
'define HLT     4'b1001

// Define Condition Code fields

'define CARRY   psr[0]
'define EVEN    psr[1]
'define PARITY  psr[2]
'define ZERO    psr[3]
'define NEG     psr[4]

// Define Condition Codes

                // Condition Code set when ...
```

```verilog
`define CCC    1        // Result has carry
`define CCE    2        // Result is even
`define CCP    3        // Result has odd parity
`define CCZ    4        // Result is Zero
`define CCN    5        // Result is Negative
`define CCA    0        // Always

`define RIGHT  0        // Rotate/Shift Right
`define LEFT   1        // Rotate/Shift Left

// Continuous assignment for queue_full

assign  queue_full = (qsize == QDEPTH) ;

// Functions for ALU operands and result
function [WIDTH-1:0] getsrc;
input [WIDTH-1:0] in ;
begin
  if (`SRCTYPE === `REGTYPE) begin
      getsrc = RFILE[`SRC] ;
  end
  else begin // immediate type
      getsrc = `SRC ;
  end
end
endfunction

function [WIDTH-1:0] getdst;
input [WIDTH-1:0] in ;
begin
  if (`DSTTYPE === `REGTYPE) begin
      getdst = RFILE[`DST] ;
  end
  else begin // immediate type
    $display("Error:Immediate data can't be destination.");
  end
end
endfunction

// Functions/tasks for Condition Codes

function checkcond;     // Returns 1 if condition code is set.
input [4:0] ccode ;
begin
  case (ccode)
   `CCC : checkcond = `CARRY ;
   `CCE : checkcond = `EVEN ;
   `CCP : checkcond = `PARITY ;
   `CCZ : checkcond = `ZERO ;
   `CCN : checkcond = `NEG ;
   `CCA : checkcond = 1 ;
  endcase
end
endfunction
```

## Modeling a Pipelined Processor

```
task clearcondcode ;      // Reset condition codes in PSR.
begin
  psr = 0 ;
end
endtask

task setcondcode ;// Compute the condition codes and set PSR.
input [WIDTH:0] res ;
begin
  'CARRY = res[WIDTH] ;
  'EVEN  = ~res[0] ;
  'PARITY = ^res ;
  'ZERO  = ~(|res) ;
  'NEG = res[WIDTH-1] ;
end
endtask

// Functions and Tasks

task fetch ;
begin
  IR_Queue[fptr] = MEM[pc] ;
  fetched = 1 ;
end
endtask

task execute ;
begin
  if (!mem_access) ir = IR_Queue[eptr] ; // New IR required?

  case ('OPCODE)
    'NOP : begin
            if (debug) $display("Nop ...") ;
          end
    'BRA : begin
            if (debug) $write("Branch ...") ;
            if (checkcond('CCODE) == 1) begin
              pc = 'DST ;
              branch_taken = 1 ;
            end
          end
    'LD : begin
          if (mem_access == 0) begin
            mem_access = 1 ; // Reserve next cycle
          end
          else begin // Mem access
            if (debug) $display("Load ...") ;
            clearcondcode ;
            if ('SRCTYPE) begin
              RFILE['DST] = 'SRC ;
            end
            else RFILE['DST] = MEM['SRC] ;
              setcondcode({1'b0,RFILE['DST]}) ;
              mem_access = 0 ;
            end
          end
```

```
    'STR : begin
            if (mem_access == 0) begin
               mem_access = 1 ; // Reserve next cycle
            end
            else begin // Mem access
               if (debug) $display("Store ...") ;
               clearcondcode ;
               if ('SRCTYPE) begin
                  MEM['DST] = 'SRC ;
               end
               else MEM['DST] = RFILE['SRC] ;
                  mem_access = 0 ;
            end
         end

         // ADD, MUL, CMP, SHF, and ROT
         // are modeled identical to the other model.

    'HLT : begin
            $display("Halt ...") ;
            halt_found = 1 ;
         end
    default : $display("Error:Wrong Opcode in instruction.");
    endcase
    if (!mem_access) executed = 1 ; // Instruction executed?
end
endtask

task write_result ;
begin
   if (('WOPCODE >= 'ADD) && ('WOPCODE < 'HLT)) begin
      if ('WDSTTYPE == 'REGTYPE) RFILE['WDST] = wresult ;
      else MEM['WDST] = wresult ;
      result_ready = 0 ;
   end
end
endtask

task flush_queue ;
begin
   // pc is already modified by branch execution
   fptr = 0 ;
   eptr = 0 ;
   qsize = 0 ;
   branch_taken = 0 ;
end
endtask

task copy_results ;
begin
   if (('OPCODE >= 'ADD) && ('OPCODE < 'HLT)) begin
      setcondcode(result) ;
      wresult = result ;
      wir = ir ;
      result_ready = 1 ;
   end
```

# CHAPTER 4

# Modeling System Blocks

In the previous chapter we saw how to model a processor at the instruction set level and its function at the behavioral level. In this chapter we present a structural model of the SISC processor and show how to model its various building blocks. We begin with the block diagram of the SISC processor and present its corresponding structural model to show the interconnections of its building blocks. In subsequent sections we develop functional models for these blocks, namely, the datapath, the memory elements, the clock generator, and the control unit.

In the datapath section we present an incrementer, an adder, a barrel shifter, and a multiplier. The section on memories includes a register file, a random-access memory, and a content-addressable memory. The clock generator section includes a single- and a two-phase clock generator. In the last section we present a state machine model of the nonpipelined version of the processor control unit.

The models shown in this chapter can be synthesized. We have not made an attempt to optimize these models of functional building blocks for any specific purpose; rather, just one way of synthesizing the blocks. A generic library consisting of some basic cells, and a small number of very simple constraints for regulating flattening, structuring, and mapping processes was used. It is important to note that these examples should be taken as guidelines and references, and you should suitably modify them to apply in your design environment.

## Structural Model

We take the pipelined SISC processor model of the previous chapter and proceed to realize the description in hardware. First, we need to identify all the data paths such as registers, counters, the register file, and the arithmetic-logic unit (ALU). Next, we need to identify all the control paths, such as programmable-logic arrays (PLAs) and random logic, that implement various state machines. Having done that, we choose these basic building blocks to construct our data and control paths, and interconnect them to achieve the desired functionality.

Most of the datapath elements are easily identified. We need memory with 4K 32-bit words, a register file with two read ports and one write port, a 12-bit register for a program counter (PC), a 32-bit instruction register (IR), a 5-bit processor status register (PSR), and a 32-bit ALU. In addition, implementing the pipeline requires the instruction queue, the registers (fptr and eptr) to indicate the status of various instructions in the pipeline, and the registers to retain the result for latter use (wir and wresult). As you can see, the declarations in the functional model form the basis for identifying the datapath elements.

The control paths are harder to identify. A simple approach is to take all the control signals required by the datapath elements for proper operation and create one control module. Such a control module can be split hierarchically into separate control blocks such as a memory control block, a branch detection block, an instruction queue management block, and so on. Each control block can be of varying complexity, from a handful of simple equations to one or more state machines that require detailed microprogramming.

A block diagram which interconnects the functional blocks of the SISC is presented in Figure 4.1.

A complete Verilog description of the SISC model created from instantiating the lower-level building blocks is shown in Figure 4.2. In order to keep it simple, we have not included the instruction queue in the model.

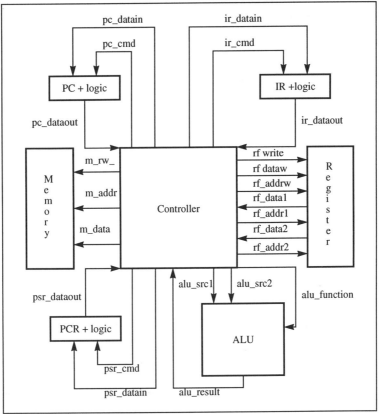

Figure 4.1 SISC block diagram

```
module system ;
parameter ADDRSIZE = 12,
          WIDTH    = 32 ;
defparam  PSR.WIDTH = 5,
          PC.WIDTH  = 12,
          IR.WIDTH  = 32 ;
wire   phase1, phase2,
       m_rw_, rf_write,
       halt, reset ;
```
**continued**

Figure 4.2 SISC Model

```
// Commands: Hold, Clear, Load, CountUp
wire [1:0]  ir_cmd, pc_cmd, psr_cmd ;

// NOP, ADD, MUL, CMP, SHF, ROT
wire [3:0]  alu_function ;

// PSR Flags - C, E, P, Z, N
wire [4:0]  psr_dataout, psr_datain ;

wire [ADDRSIZE-1:0]
       pc_addrout,
       pc_addrin,
       m_addr,
       rf_addrw,
       rf_addr1,
       rf_addr2 ;
wire [WIDTH-1:0] m_data,
       ir_dataout,
       ir_datain,
       rf_dataw,
       rf_data1,
       rf_data2,
       alu_src1,
       alu_src2 ;
wire [WIDTH:0] alu_result ;

// Instantiate predefined building blocks

clock    CLK (phase1, phase2) ;
memory   MEM (phase1, phase2, m_data, m_addr, m_rw_) ;
regfile  RFILE(phase1, phase2, rf_dataw,
           rf_addrw, rf_write,
           rf_data1, rf_addr1, rf_data2, rf_addr2) ;
regcntr  PC  (phase1, phase2, pc_addrout,
           pc_addrin, pc_cmd, reset) ;
regcntr  IR  (phase1, phase2, ir_dataout,
           ir_datain, ir_cmd, reset) ;
regcntr  PSR (phase1, phase2, psr_dataout,
           psr_datain, psr_cmd, reset) ;
alu      ALU (phase1, phase2, alu_result,
           alu_function, alu_src1, alu_src2) ;
cntrl    CONTROLLER (phase1, phase2,
           halt, reset, m_data, m_addr, m_rw_,
           rf_dataw, rf_addrw, rf_write,
           rf_data1, rf_addr1, rf_data2, rf_addr2,
           pc_addrout, pc_addrin, pc_cmd,
           ir_dataout, ir_datain, ir_cmd,
           psr_dataout, psr_datain, psr_cmd,
           alu_result, alu_function, alu_src1,alu_src2);
endmodule
```

Figure 4.2  Structural model of the processor (continued)

## Datapath

The arithmetic-logic unit of a typical processor performs all the arithmetic, logic, and shift/rotate functions. The arithmetic functions generally include addition, subtraction, multiplication, division, increment, and possibly others. The logic functions consist of AND, OR, NOT, and so forth. The special functions include shift, rotate, and comparison operations. The overall functionality of the ALU, however, is determined by the instruction set as defined by its architecture.

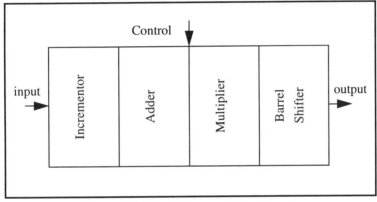

Figure 4.3 Block diagram of the datapath

Referring to Figure 4.3, the datapath of the SISC processor is required to implement the following instructions:

```
ADD,  SUB,  MUL,  CMP    (arithmetic)
SHF,  ROT                (shift, rotate)
```

A block diagram of the datapath, as it is normally laid out in a VLSI processor design, is shown in Figure 4.3. One side of the datapath receives all the data inputs while the opposite side provides all the data outputs. The control signals are orthogonal to the flow of data in the datapath.

The input to the datapath consists of two 32-bit data buses. Some instructions require only one data bus while other instructions use both the buses. The datapath blocks necessary to implement the SISC processor instructions are an adder, a multiplier, and a barrel shifter. The incrementer block of the datapath is used to update the value of the program counter.

In the following sections we will model each of these functional blocks of the datapath.

```
module increment (in, control, out,
                Nflag, Pflag, Eflag, Zflag, Cflag);
input   [31:0] in;    // input operand
input   control;      // increment or pass
output  [31:0] out;   // result operand
output  Eflag, Zflag, Cflag, Nflag, Pflag;

parameter LOAD  = 0 ,
          INCR  = 1 ,
          TRUE  = 1 ,
          FALSE = 0 ;

  wire [31:0] out -
           (control === INCR) ? (in + 1) : in;

  wire Cflag =
           (~(|out) && control)  ? TRUE  : FALSE;

  //  set condition codes -- remaining flags
  set_flags  U1(out, Nflag, Pflag, Eflag, Zflag);

endmodule
```

Figure 4.4  Model of the incrementer

**Incrementer**

Figure 4.4 gives the model of the incrementer. It implements the incrementer function as a conditional operator which either increments the input or transfers the value unmodified to the output, depending on the value of the control signal. The carry flag is set if the control increments the input and if the output is zero.

A decrement function can be obtained by making minor modifications to the incrementor model as follows:

```
wire [31:0] out =
            (control === INCR) ? (in + 1) :
            (control === DECR) ? (in - 1) : in;
```

# Modeling System Blocks

This, however, requires that the control signal be 2 bits wide. If the output is connected to a data bus, shared by several other drivers, the 2 bit control signal may be encoded with four functions - LOAD, INCR, DECR, TRI. The new function, TRI, would tristate the outputs to avoid conflict with the driver of the bus. The set_flags module sets various condition codes in the processor status register.

**Adder**

Figure 4.5 represents a high-level abstraction of the adder behavior. The adder implements the ADD and SUB instructions of the datapath. The control signal selects between addition and subtraction in the conditional expression. Various condition codes are set by the set_flags module.

The adder described in Figure 4.5 uses only the "+" operator and does not provide an insight into the internals of its implementation. The carry-look-ahead adder of Figure 4.6 describes a model of what may be an inefficient carry-look-ahead adder. This model uses continuous assignments to generate the 32-bit carry-generate and carry-propagate signals. A for loop is used to calculate the ripple-carry chain where each bit of the carry chain depends upon its associated carry-generate, carry-propagate, and the carry output from the previous bit.

```
module adder (operand1,operand2,result,control,
              Cflag, Pflag, Eflag, Zflag, Nflag);

input   [31:0] operand1, operand2;
input   control;
output  [31:0] result;
output  Cflag, Pflag, Eflag, Zflag, Nflag;

parameter ADD  = 0 ,
          SUB  = 1 ;

  assign {Cflag, result} = (control === ADD) ?
                            (operand1 + operand2)  :
                            (operand1 - operand2)  ;

  //  set condition codes -- remaining flags
  set_flags  U2(result, Nflag, Pflag, Eflag, Zflag);

endmodule
```

Figure 4.5 Model of the adder

```
module adder_c (operand1,operand2,cin,result,cout);

input   [31:0]  operand1, operand2 ;
input   cin ;
output  [31:0]  result ;
output  cout        ;

reg     [31:0]  carrychain;

// Generate carry
wire    [31:0]  g = operand1 & operand2 ;

// Propagate carry
wire    [31:0]  p = operand1 ^ operand2 ;

always @(operand1 or operand2 or cin)
  begin :carry_generation
    integer i;
    carrychain[0] = g[0] + (p[0] & cin );
    for (i = 1; i <= 31; i = i + 1) begin
      #0 // force evaluation
      carrychain[i] = g[i] + (p[i] & carrychain[i-1]);
    end
  end

wire [32:0] shiftedcarry = {carrychain, cin} ;

// Compute the sum
wire [31:0] result = p ^ shiftedcarry;

// Carry out
wire    cout = shiftedcarry[32];

endmodule
```

Figure 4.6 Model of the carry-look-ahead adder

In an actual implementation of a 32-bit adder, a 32-bit ripple chain would not be used because of the long propagation delay of the carry. Instead, it might use multiple levels of 8-bit ripple-carry adders with carry-look-ahead. Using the technique described here, models can be developed which correspond to the actual implementation.

**Barrel Shifter**

The shift and rotate (SHF, ROT) instructions of the datapath are modeled by the barrel shifter of Figure 4.7. The barrel shifter models the two rotate and two shift functions: shift-left, shift-right, rotate-left, and rotate-right. The rotation is modeled by repeated shift operations.

```
module barrel_shifter (in, direction, type, count,
            result, Nflag, Pflag, Eflag, Zflag);

input   [31:0]  in;       // input operand
input   direction;        // left or right
input   type;             // shift or rotate
input   [3:0]   count;    // shift count
output  [31:0]  result;   // result operand
output  Nflag, Pflag, Eflag, Zflag;

parameter LEFT   = 0 ,
          RIGHT  = 1 ,
          SHIFT  = 0 ,
          ROTATE = 1 ;

// Invoke the shift/rotate functions
// to obtain the result.

wire [31:0] result = (type === ROTATE) ?
        (rotate(in , count, direction)) :
        (shift (in , count, direction)) ;

// This function implements the SHF instruction
function [31:0] shift;

  input [31:0] in;
  input [3:0]  count;
  input direction;

  begin
    shift = (direction === RIGHT) ?
        (in >> count) : (in << count);
  end

endfunction

// This function implements the ROT instruction
function [31:0] rotate;

  input [31:0] in;
  input [3:0]  count;
  input direction;

  reg    [31:0] reg_rotate;
  reg    t;
  integer i;                              continued
```

Figure 4.7 Model of the barrel-shifter

```
begin
  reg_rotate = in[31:0];
  if (direction === RIGHT)
    for (i = 0; i < count; i = i + 1) begin
      t = reg_rotate[0];
      reg_rotate[30:0] = reg_rotate[31:1];
      reg_rotate[31] = t;
    end
  else if(direction === LEFT)
    for (i = 0; i < count; i = i + 1) begin
      t = reg_rotate[31];
      reg_rotate[31:1] = reg_rotate[30:0];
      reg_rotate[0] = t;
    end
  rotate = reg_rotate;
end
endfunction

// set condition codes -- remaining flags
set_flags U3(result,Nflag,Pflag,Eflag,Zflag);

endmodule
```

Figure 4.7 Model of the barrel-shifter (continued)

**Multiplier**

The multiplication of two data operands in Verilog can be simply written by using the "*" arithmetic operator. Figure 4.8 models the multiplier which implements the MUL instruction of the datapath. Both the operands of the multiplier are 16-bit wide so as to obtain a 32-bit result. Having 32-bit operands would require the processor either to have a 64-bit data bus or to multiplex the data in two consecutive cycles. This would increase the complexity of the instruction set and require the processor to support multiple cycle operations.

**Setting Condition Codes**

In modeling the datapath, we saw that each module needs to set the flags for individual condition codes of the processor status register, PSR. Instead of replicating the code in each of the modules, the code is encapsulated in a single module (set_flags) and is instantiated by the individual modules of the datapath.

## Modeling System Blocks

In the set_flags module of Figure 4.9, the negative flag is the value of the most significant bit of the data value. The parity flag checks for an odd parity and can be derived by the reduction exclusive-OR of the data. The even flag is set high if the least significant bit of the data is zero. The zero flag is set high if all the bits of the data are logically low, and this flag is obtained by negating the reduction-OR of the data.

```
module multiplier (operand1,operand2,result,
         Cflag, Pflag, Eflag, Zflag, Nflag);
  input   [15:0] operand1, operand2;
  output  [31:0] result;
  output  Cflag, Pflag, Eflag, Zflag, Nflag;

  assign {Cflag,result} = operand1 * operand2;

  set_flags  U2 (result,Nflag,Pflag,Eflag,Zflag);

endmodule
```

Figure 4.8  Model of the multiplier

```
module set_flags (value,
                 Nflag, Pflag, Eflag, Zflag);

  input   [31:0] value;
  output  Nflag, Pflag, Eflag, Zflag;

  wire   Nflag = value[31];
  wire   Pflag = ^(value[31:0]);
  wire   Eflag = ~value[0];
  wire   Zflag = ~(|(value[31:0]));

endmodule
```

Figure 4.9  Model for setting condition codes

## Memories

In this section we will show how to model memory elements such as a random-access memory, a content-addressable memory, and a register file. The latter is not used in our processor design but is provided here as an example.

Typically, design and layout of RAMs and Register Files are done manually or generated automatically through other special purpose tools. This is primarily due to regular and highly compact structure, and very little scope for optimization in the control logic. While synthesizing a design with RAMs and Register Files, the designer instructs the tool explicitly to generate a black box representation or nothing at all. Later, the designer manually combines previously designed memory and other such special modules with the synthesized model of the remaining design.

**Random-Access Memory**

Read/write operations on the random-access memory (RAM) are controlled by the clock phases. In a two-phase design, one phase is used for the read cycle and the other is used for the write cycle.

Figure 4.10 shows the model for the RAM used in the SISC processor design. The write cycle of the RAM is controlled by the ph2 clock and the write enable signal wrenable. The do_write block is set to trigger when either of the two signals change. We assume that the address or data can change anytime during the phase as long as these signals maintain a setup and hold time before the trailing edge of the phase clock. These signals are included in the specify block for the module. In the model of RAM, the read is a continuous assignment where the output data register is assigned the value of the data at the address currently on the address bus, as long as the phase ph1 is true.

The read and write phases can be switched depending on the design requirements. The data needs to be valid only for a predefined setup time. This is illustrated by defining setup in the module definition of the RAM. The access time is modeled as a read-delay to the output.

**Content-Addressable Memory**

In processor designs, content-addressable memories (CAMs) are frequently used to perform parallel search and comparison operations. However, CAMs need more logic in their design than RAMs, giving rise to increased area and higher cost. Our SISC processor does not use any CAM.

# Modeling System Blocks

```
'timescale 1ns / 1ps
module ram (ph1, ph2, addr,
            wrenable, datain, dataout);

  parameter WORDS       = 4095 ,
            ACCESS_TIME = 5 ;

  input  ph1, ph2, wrenable;
  input  [11:0] addr;     // address bus
  input  [31:0] datain;   // input data bus
  output [31:0] dataout;  // output data bus

  // Define RAM as a register array
  reg    [31:0] ram_data[WORDS:0];

  // Read cycle
  wire [31:0] data = (ph1 === 1) ?
           ram_data[addr[11:0]] : 32'hz;
  wire [31:0] #ACCESS_TIME dataout = data;

  // Write cycle
  always @(posedge ph2 or addr or
           datain or wrenable) begin : do_write
    if (ph2 === 1 && wrenable === 1)
       ram_data[addr[11:0]] = datain;
  end

  specify
    specparam tSetup = 0.75:1.0:1.25,
              tHold  = 0.25:0.5:0.75;
       $setup (datain, ph2, tSetup);
       $hold  (ph2, datain, tHold);
  end
endmodule
```

Figure 4.10    Model of the RAM

The RAM model of Figure 4.10 can be modified to perform CAM-like operations by using comparison functions as illustrated in Figure 4.11. The resulting output can be synchronized to one of the phase clocks as needed.

**Register File**

The SISC processor uses a set of sixteen general-purpose registers. The architecture of the processor requires two read operations on the registers in a cycle but only one write operation. This implies that the register set can be modeled as a multiport RAM.

The model of a register file with two read ports and one write port is shown in Figure 4.13. As in the model of the RAM, the read and write cycles consist of two loops executing in parallel, one loop for read and one loop for write. During the read cycle, the dataout output registers get the value in the RAM at the address determined by the two address buses. The address bus Aaddr is also used during the write cycle and is multiplexed by the write enable signal.

```
module cam_ram (comparand,mask,result,.....);

....................
output [WORDS:0] result;
input  [31:0] comparand, mask;
reg    [31:0] comparand, mask;
reg    [31:0] ram_data[WORDS:0];
                // CAM as a register array

// associative operation
wire [WORDS:0] result =
   {((cam_data[WORDS] & mask) === comparand) ?
                          1'h1 : 1'h0,
    ((cam_data[WORDS-1] & mask) === comparand) ?
                          1'h1 : 1'h0,
    ................
    ................
((cam_data[0] & mask) === comparand) ?
                          1'h1 : 1'h0};
................
................

endmodule
```

Figure 4.11  Model of a CAM

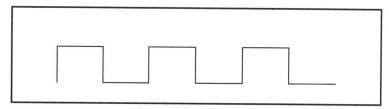

Figure 4.12  A 50% duty cycle clock

Modeling System Blocks

```
'timescale 1ns / 1ps
module reg_file (ph1, ph2, Aaddr, Baddr,
      wrenable, datain, Adataout, Bdataout);

input   ph1, ph2;    // clock phases
input   wrenable;    // write-enable
input   [3:0]  Aaddr, Baddr;  // address
input   [31:0] datain;         // input data
output  [31:0] Adataout, Bdataout; // output data buses
reg     [31:0] Aregout, Bregout; // holds data values
reg     [31:0] ram_data[15:0];    // RAM register array

parameter  ACCESS_TIME = 5 ;

// read cycle
wire [31:0] #ACCESS_TIME Adataout = Aregout;
wire [31:0] #ACCESS_TIME Bdataout = Bregout;
always @(posedge ph1 or Aaddr or Baddr or wrenable)
if (ph1 === 1 && wrenable === 0) begin
  Aregout = ram_data[Aaddr[3:0]];
  Bregout = ram_data[Baddr[3:0]];
end

// write cycle
always @(posedge ph2 or Aaddr or datain or wrenable)
if (ph2 === 1 && wrenable === 1)
  ram_data[Aaddr[11:0]] = datain;

specify
  specparam tSetup = 1.0, tHold = 0.5;
  $setup (datain, ph2, tSetup);
  $hold  (ph2, datain, tHold);
endspecify
endmodule
```

Figure 4.13  Model of the register file

## Clock Generator

The clock signal synchronizes events in different parts of the circuit and, in particular, synchronizes the updating of memory elements, such as flip-flops and latches. In most cases the master clock is generated outside the VLSI chip. The on-chip clock generator is coupled to the master clock through a phase-locked loop which provides stability to the clock by compensating for variations in the duty cycle. Although our processor uses a two-phase clock, we will present models for both single- and two-phase clocks.

```
module clock_gen (masterclk);

  parameter MASTERCLK_PERIOD = 10 ;

  output masterclk;
  reg    masterclk;

  initial
    masterclk = 0;

  // oscillation at a given period
  always begin
    # MASTERCLK_PERIOD/2
    masterclk = ~masterclk;
  end

endmodule
```

Figure 4.14  Model of the 50% duty cycle clock

**Single-Phase Clock**

The simplest clock model generates a single 50% duty cycle as shown in Figure 4.12.

Figure 4.14 describes the model for the simple clock generator. It defines a register, masterclk, which switches state from 1 to 0 and from 0 to 1 at a given frequency. The period of the clock depends on the design and is user defined. A more general clock generator has a variable duty cycle.

**Two-Phase Clock**

While the clock generator of Figure 4.15 is sufficient for single-phase designs, most current processors are based on two-phase clocks. In this scheme, the basic memory element is a transparent latch whose output follows its input as long as the clock is high but stays latched when the cycle goes low.

In a two-phase clock design, some latches are clocked by phase-1 of the clock, and some are clocked by phase-2. Usually the two types of latches alternate; the output of phase-1 latches feed (possibly through some combinational logic) the data input of phase-2 latches, and vice versa.

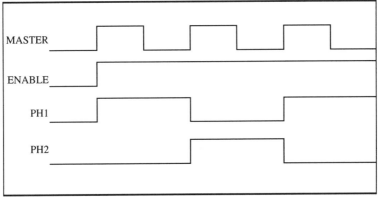

Figure 4.15  A two-phase, nonoverlapping clock

Figure 4.17 provides a model for generating a two-phase clock. The phase clocks are one-half the frequency of the master clock and are synchronized to the leading edge of the master clock. The phase generator produces two-phase, nonoverlapping clocks at one-half the frequency of the master clock.

```
module clock_driver;

reg enable;

clock_gen P1(master);
phase_gen P2(master, enable, ph1, ph2);

initial begin
  enable = 0;
  #10 enable = 1;
  #100 enable = 0;
  #100
  $stop;
end

initial begin
  $gr_waves("MASTER", master,
   "ENABLE", P2.enable,
   "PH1", P2.ph1,
   "PH2", P2.ph2);
end

endmodule
```

Figure 4.16  Driver for the clock generator

```
module phase_gen (masterclk,enable,ph1,ph2);
parameter NON_OVERLAP = 1 ;
input    masterclk, enable;
output   ph1, ph2;
reg      ph1, ph2, reset;
initial begin // reset all signals
  ph1 = 0;
  ph2 = 0;
  reset = 0;
end

always @(posedge masterclk) begin : generate_phases
  if (enable == 1) begin
    ph2 = 0;
    # NON_OVERLAP
    ph1 = 1;
    @(posedge masterclk) begin
      ph1 = 0;
      # NON_OVERLAP;
      ph2 = 1;
    end
  end
end

always @(posedge masterclk) begin
  if (enable == 0) begin
    reset = 1;
    @(posedge masterclk) begin
      if (enable === 0 && reset === 1) begin
        ph1 = 0;
        ph2 = 0;
        reset = 0;
      end else
        reset = 0;
    end
  end
end

endmodule
```

Figure 4.17  Model of the two-phase clock generator

In the phase generator module, phase_gen, a reset mechanism is provided in which the phase clocks are shut down if the enable signal is low for two consecutive rising edges of the master clock. Figure 4.18 shows the waveforms obtained by combining the two models together and running them as one clock unit.

Modeling System Blocks

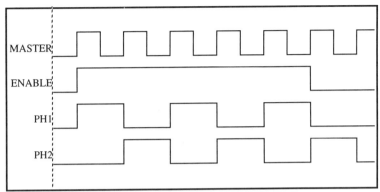

Figure 4.18 Output of the two-phase clock generator

```
module control_pla (pc_dataout, pc_datain, pc_cmd,
       ir_datain, ir_cmd, ir_dataout, rf_write,
       rf_dataw, rf_addrw, rf_data1, rf_addr1,
       rf_data2, rf_addr2, alu_function,alu_src1,
       alu_src2, alu_result, psr_datain, psr_cmd,
       psr_dataout, m_data, m_addr, m_rw_ , state);

'define      NOP        4'h0
'define      BRA        4'h1
'define      LOAD       4'h2
'define      STORE      4'h3
'define      ADD        4'h4
'define      MUL        4'h5
'define      CMP        4'h6
'define      SHF        4'h7
'define      ROT        4'h8
'define      HALT       4'h9
'define      FETCH      2'h1
'define      EXECUTE    2'h2
'define      WRITE      2'h3

input    [11:0] pc_dataout;
input    [1:0]  state;
input    [4:0]  psr_dataout;
input    [31:0] ir_dataout, rf_data1, rf_data2;
input    [32:0] alu_result;
inout    [31:0] m_data;
reg      [31:0] m_data_reg;
output          rf_write, m_rw_;
output   [1:0]  pc_cmd, ir_cmd, psr_cmd;
output   [3:0]  rf_addrw, rf_addr1, rf_addr2, alu_function;
                                                continued
```

Figure 4.19  Model of the control-unit PLA

```
output [4:0]    psr_datain;
output [11:0]   pc_datain, m_addr;
output [31:0]   ir_datain, rf_dataw, alu_src1, alu_src2;
'define     CLEAR_REG    2'h0
'define     HOLD_VAL     2'h1
'define     LOAD_REG     2'h2
'define     COUNTUP      2'h3
'define     IR_OPCODE    ir_dataout[31:28]
'define     IR_SRC_TYP   ir_dataout[27]
'define     IR_DST_TYP   ir_dataout[26]
'define     IR_SRC       ir_dataout[23:12]
'define     IR_SRC_REG   ir_dataout[15:12]
'define     IR_DST       ir_dataout[11:0]
'define     IR_DST_REG   ir_dataout[3:0]
'define     IMMEDIATE    1
'define     MEM_READ     1
'define     MEM_WRITE    0
'define     REG_READ     0
'define     REG_WRITE    1
'define     IR_CCODES    ir_dataout[27:23]

wire [31:0] m_data = m_data_reg;
wire m_rw_ = ('IR_OPCODE === 'STORE && state === 'WRITE) ?
             'MEM_WRITE : 'MEM_READ;
wire [11:0] m_addr = (state === 'FETCH) ? pc_dataout :
             (('IR_OPCODE === 'STORE) ? 'IR_DST : 'IR_SRC);

always @ (state) begin
   if(state === 'EXECUTE || state === 'WRITE) begin
      if('IR_OPCODE === 'STORE && 'IR_SRC_TYP !=
'IMMEDIATE)
         m_data_reg = rf_data1;
      else
      if('IR_OPCODE === 'STORE && 'IR_SRC_TYP ===
'IMMEDIATE)
         m_data_reg = 'IR_SRC ;
      else m_data_reg = 32'hz;
   end
   else
      m_data_reg = 32'hz;
end

wire [3:0] alu_function =
      ('IR_OPCODE >= 'ADD &&
       'IR_OPCODE <= 'ROT &&
       state === 'EXECUTE) ? 'IR_OPCODE : 'NOP;
wire [31:0] alu_src1 = ('IR_SRC_TYP != 'IMMEDIATE) ?
       (('IR_OPCODE != 'ROT &&
         'IR_OPCODE != 'SHF) ? rf_data1 : 'IR_SRC) :
       'IR_SRC;
```
**continued**

Figure 4.19   Model of the control-unit PLA(continued)

# Modeling System Blocks

```verilog
wire [31:0] alu_src2 = rf_data2;
wire [3:0]  rf_addr1 = 'IR_SRC_REG ;
wire [3:0]  rf_addr2 = 'IR_DST_REG ;
wire [3:0]  rf_addrw = 'IR_DST_REG ;
wire rf_write = ('IR_OPCODE != 'BRA &&
         'IR_OPCODE != 'STORE &&
         'IR_OPCODE != 'NOP &&
         'IR_OPCODE != 'HALT &&
          state === 'WRITE) ? 'REG_WRITE : 'REG_READ;
wire [31:0] rf_dataw =
         (rf_write &&
         'IR_OPCODE === 'LOAD &&
         'IR_SRC_TYP != 'IMMEDIATE) ? m_data :
         ((rf_write && 'IR_OPCODE === 'LOAD &&
         'IR_SRC_TYP === 'IMMEDIATE) ? 'IR_SRC :
alu_result);
wire [1:0] ir_cmd = (state === 'FETCH) ?
         'LOAD_REG : ((state === 'EXECUTE) ? 'HOLD_VAL :
         (('IR_OPCODE === 'HALT) ? 'CLEAR_REG :
'HOLD_VAL));
wire [31:0] ir_datain = (ir_cmd === 'LOAD_REG) ?
         m_data : ((ir_cmd === 'CLEAR_REG) ?
         32'h0 : ir_dataout);
wire branch_taken = | ( 'IR_CCODES & psr_dataout);
wire [1:0] pc_cmd = (state === 'WRITE) ?
         (('IR_OPCODE === 'HALT) ?
         'CLEAR_REG : (branch_taken ?
         'LOAD_REG : 'COUNTUP)) : 'HOLD_VAL;
wire [11:0] pc_datain = (state == 'WRITE) ?
         (('IR_OPCODE == 'HALT) ? 12'h0 : (branch_taken ?
         'IR_DST : (pc_dataout + 1))) : pc_dataout;
wire [1:0] psr_cmd = ('IR_OPCODE >= 'ADD &&
         'IR_OPCODE <= 'ROT &&
         state === 'WRITE) ? 'LOAD_REG :
         (('IR_OPCODE === 'HALT) ? 'CLEAR_REG : 'HOLD_VAL)
;
wire zflag = ~(|alu_result[31:0]);
wire eflag = ~(alu_result[0]);
wire pflag = ^(alu_result[31:0]);
wire [4:0] psr_datain = (psr_cmd === 'CLEAR_REG) ? 5'h0
         ((psr_cmd === 'LOAD_REG) ?
         {alu_result[31], zflag, pflag, eflag,
alu_result[32]}
         : psr_dataout);

endmodule
```

Figure 4.19  Model of the control-unit PLA (continued)

**Clock Driver**

The clock_driver module invokes the clock_gen and phase_gen modules to create a two-phase, nonoverlapping clock waveform. The clock_driver module is shown in Figure 4.16.

## Control Unit

The control unit coordinates the operations among the various functional blocks of the processor. Figure 4.19 gives the equations of all the control lines for the nonpipelined processor. Similar equations can be developed for the pipelined version.

The output of each control signal depends on the state of the processor. The processor goes through three states—fetch, execute, and write—in sequence for each instruction.

The memory read-write signal, m_rw, is a write signal only if the state is write and the instruction opcode is a write. During the fetch cycle, the memory address m_addr is set to the program counter in order to fetch the next instruction. During the other states, m_addr gets its value from the instruction register source or destination fields.

The memory data bus, m_data, is a bidirectional bus which gets its value from m_data_reg. In fetch state, m_data acts as an input and, therefore, m_data_reg is tristated. In nonfetch state, if the opcode is a store, the register gets its value from either the register file or the instruction register depending on the type of the instruction.

In execute state, the alu_function signal is taken from the operand field of the IR, while the operands are taken from the register file or from the IR (for an immediate operand).

The data input to the register file is normally taken from the output of the ALU. The only exception is during execution of a load instruction. In that case, the input to the register file comes from the instruction register for immediate instruction or from the memory data register for nonimmediate instruction.

The operation of the program counter, the instruction register, and the status register depends on their respective command registers (pc_cmd, ir_cmd, and psr_cmd). During a given state, a command register can instruct a register to retain its value, to load a new value in, to increment its value, or to clear the register. In the fetch state, the instruction register is updated with a new value from memory and maintains its value during all other states. The program counter increments its value at the end of every write state except when a branch occurs. The value of the status register depends on the output of the ALU and is derived as a function of alu_result. All three registers—PC, IR, and PSR—are cleared when the opcode is HLT.

The control block as described above models only the combinational part of the control block. The logic for generating the timing signals is not modeled. This logic has to implement a state machine to synchronize the updating of the control block state (IR, PC, and so forth) to the right processor state (fetch, execute, write) and to the right clock phase.

## Summary

In this chapter we presented a structural model of the SISC, based upon its instruction set model from the previous chapter. Various building blocks of a computer and a central processor were modeled using the SISC as an example. The equations for the control unit of the CPU were also developed. These can be implemented using a PLA or some library components. These models can be used as guidelines to design more complex architectures.

CHAPTER

# 5

# Modeling Cache Memories

In this chapter we examine the process of designing a simple cache system in Verilog HDL. The description can be synthesized to obtain a gate level implementation. At the end of this chapter we nsider ways to improve the basic design.

A cache in a computer system is a small, fast, local memory that stores data from the most frequently accessed addresses. The cache is always small compared to main memory because cache RAMs are more expensive than the slower dynamic RAMs used for main memory. As a result, only a small portion of the main memory can be stored in the cache. The efficiency of a cache is measured by the cache hit ratio, i.e., the number of times data is accessed from the cache over the total number of memory accesses. Typical hit ratios are in the range of eighty to one hundred percent.

First, we examine how the cache communicates with the environment around it. Then we discuss cache architecture followed by a synthesized implementation of a cache model. We also give suggestions of methods for testing and improving it.

# Interfaces

A cache system typically lies between the processor and the main system bus. The following block diagram (Figure 5.1) describes the signals and buses that the cache needs to communicate with the processor and the system. Note that the system bus is synonymous with main memory, and the terms will be used interchangeably in the text.

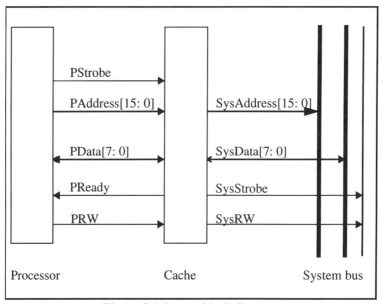

**Figure 5.1** System block diagram

**Processor Interface**

The processor interface consists of the processor address bus, PAddress[15: 0], the processor data bus, PData[7: 0], and control signals PRW, PStrobe and PReady. PStrobe is asserted when the processor is starting a bus transaction and a valid address is on the PAddress bus. PReady is used to signal to the processor that the bus transaction is completed. The timing diagram in Figure 5.2 demonstrates a simple read cycle. The PRW signal is high for a read and low for a write.

# Modeling Cache Memories

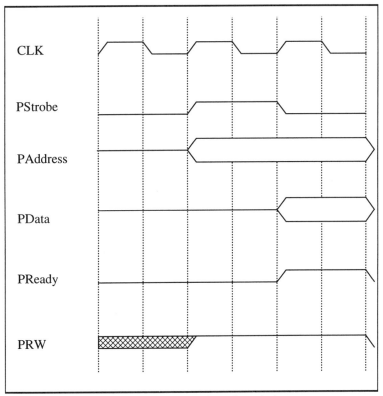

**Figure 5.2** Processor interface timing

**System Bus Interface**

For our cache, we assume a simple bus model. For a read operation, SysAddress is first presented to the bus along with the SysStrobe signal and the SysRW signal. The SysRW signal is high for read operations and low for write operations. After a set number of wait states, the data is returned. A write operation is similar, but the data is driven onto the PData bus immediately by the Cache and then is held a number of wait states before another write operation is issued. Figure 5.3. shows a system read operation with one wait state.

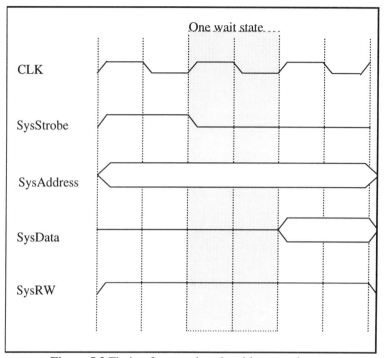

**Figure 5.3** Timing for a read cycle with one wait state

## Cache Architecture

To demonstrate general design techniques, the simplest of all cache architectures, a direct-mapped cache suffices. A direct mapped cache consists of a single tag RAM, data RAM, and a controller.

We will assume that the processor has a 16-bit address which gives it an address space of 64K bytes. With this in mind, a cache size of 1K bytes is chosen as a reasonable trade-off between cost and performance for this processor. The appropriate size for a cache is directly related to the hardware cost and to the required performance for the type of code to be run on the processor. In practice, this can be determined by analyzing execution traces of typical programs.

# Modeling Cache Memories

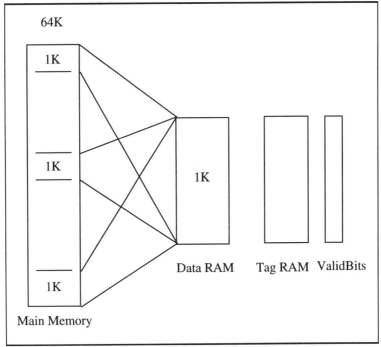

**Figure 5.4** Mapping between main memory and cache

The main memory is divided up into 64 blocks of 1K each. The mapping between the cache memory and the main memory is shown in Figure 5.4. Note that each location in the cache is mapped to 64 different locations in main memory. For example, locations 52, 1K+52, 2K+52, ... , 63K+52 in main memory, all map to location 52 in the cache. This means that only the least significant 10 bits of the address are needed to address the cache.

In order to identify the full address of a cache entry, each entry has a 6-bit tag that matches the 6 most significant bits of the main memory address corresponding to the entry currently stored in that cache location. The 16-bit address is divided into index and tag fields as shown in Figure

5.5. In addition, each cache entry has a single bit that indicates whether the entry is valid. Initially, all entries are set to invalid.

**Figure 5.5** Address fields

To further simplify the cache, a write-through mode of operation is used. This means that all writes from the processor update both the cache and the main memory. Thus the cache memory is always kept consistent with main memory. Cache systems may have other write policies that improve performance incrementally and add to the complexity of the cache system. The write-back approach is duscussed later in this chapter.

## Modeling Component Blocks

The success of a design depends heavily on the correct choice of partitioning. This is best demonstrated with a block diagram that shows the main functional units and the connections between them. From the block diagram we can identify sections of the design that can be synthesized. Synthesis tools excel at generating random logic and can often produce results that have better global optimization than logic that has been generated by hand. Most standard cell type designs will contain blocks like RAMs, ROMs, PLAs, datapaths, and other similar functional blocks that will not be synthesized. It is important to keep these blocks separate from those that will be synthesized.

The cache controller consists of
- Tag RAM
- Data RAM
- ValidBits RAM
- Tag comparator
- Data Multiplexors
- Control Section

# Modeling Cache Memories

The block diagram of the cache system is shown in Figure 5.6.

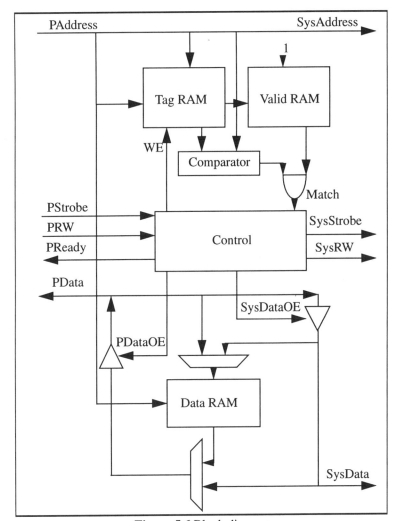

**Figure 5.6** Block diagram

**Top Level Module**

The top level module interconnects all submodules together, the sections that will be synthesized along with the sections that will require custom layout or some other form of special treatment. For clarity, the code should contain no more than the instantiations of each of the blocks and the wires that connect them together. Note the use of port connections by name. This feature of Verilog eliminates the need to remember the order of the submodule ports.

Bidirectional buses need to be defined in a slightly different way than unidirectional buses. As an example let us look at the processor data bus (PData). The following piece of code is necessary to allow the processor as well as the cache system to drive the bus at different times.

```
wire    PDataOE;
wire    ['DATA] PDataOut;
wire    ['DATA] PData=PDataOE ? PDataOut :
                 'DATAWIDTH'bz;
```

The bus `PDataOut` carries the data that the cache chip will drive onto the `PData` bus when the output enable `PDataOE` is active. At other times the cache chip drives `'bz` (high impedance) so that the final value of the `PData` bus is determined by the value of other drivers on that bus, i.e. the processor.

The code for the top level module is shown in Figure 5.7. Each of the blocks will be defined in the following sections.

```
'define READ 1
'define WRITE 0
'define CACHESIZE 1024
'define WAITSTATES 2
// number of wait states required for system accesses
'define ADDR 15:0
'define INDEX 9:0
'define TAG 15:10
'define DATAWIDTH 32
'define DATA 'DATAWIDTH-1:0
'define PRESENT 1
'define ABSENT !'PRESENT                    continued
```

**Figure 5.7** Cache: Top Level Module

```
module cache(
      PStrobe,
      PAddress,
      PData,
      PRW,
      PReady,

      SysStrobe,
      SysAddress,
      SysData,
      SysRW,

      Reset,
      Clk
      );

input            PStrobe;
input   ['ADDR]  PAddress;
inout   ['DATA]  PData;
input            PRW;
output           PReady;

output           SysStrobe;
output  ['ADDR]  SysAddress;
inout   ['DATA]  SysData;
output           SysRW;
input            Reset;
input            Clk;

// Bidirectional buses
wire             PDataOE;
wire             SysDataOE;
wire    ['DATA]  PDataOut;
wire    ['DATA]  PData=PDataOE ? PDataOut : 'bz;
wire    ['DATA]  SysData=SysDataOE ?
             PData : 'bz;
wire    ['ADDR]  SysAddress = PAddress;
wire    ['TAG]   TagRamTag;
TagRam TagRam(
      .Address   (PAddress['INDEX]),
      .TagIn     (PAddress['TAG]),
      .TagOut    (TagRamTag['TAG]),
      .Write     (Write),
      .Clk       (Clk)
                                    continued
```

**Figure 5.8** Cache: Top Level Module (continued)

```
            ValidRam ValidRam(
                    .Address     (PAddress['INDEX]),
                    .ValidIn     (1'b1),
                    .ValidOut    (Valid),
                    .Write       (Write),
                    .Reset       (Reset),
                    .Clk         (Clk));
            wire    ['DATA]DataRamDataOut;
            wire    ['DATA]DataRamDataIn;

            DataMux CacheDataInputMux(
                    .S(CacheDataSelect),
                    .A(SysData),
                    .B(PData),
                    .Z(DataRamDataIn);

            DataMux PDataMux(
                    .S(PDataSelect),
                    .A(SysData),
                    .B(DataRamDataOut),
                    .Z(PDataOut)
                    );

            DataRam DataRam(
                    .Address(PAddress['INDEX]),
                    .DataIn(DataRamDataIn),
                    .DataOut(DataRamDataOut),
                    .Write(Write),
                    .Clk(Clk)
                    );

            Comparator Comparator(
                    .Tag1(PAddress['TAG]),
                    .Tag2(TagRamTag),
                    .Match(Match)
                    );

            Control Control(
                    .PStrobe(PStrobe),
                    .PRW(PRW),
                    .PReady(PReady),
                    .Match(Match),
                    .Valid (Valid),
                    .CacheDataSelect(CacheDataSelect),
                    .PDataSelect(PDataSelect),
                    .SysDataOE(SysDataOE),              continued
```

**Figure 5.9** Cache: Top Level Module (continued)

# Modeling Cache Memories

```
            .Write           (Write),
            .PDataOE         (PDataOE),
            .SysStrobe       (SysStrobe),
            .SysRW           (SysRW),
            .Reset           (Reset),
            .Clk             (Clk)
            );
endmodule
```

**Figure 5.10** Cache: Top Level Module (continued)

## Tag RAM Module

The tag RAM module consists of a memory array capable of holding one tag for each index location, i.e., 1K x 6. The RAM is written during the low phase of CLK if the Write signal is active. The RAM is read during the high phase of CLK. The Verilog code in Figure 5.11 shows how to model the tag RAM. Note that this module is not synthesizable.

```
module TagRam(Address, TagIn, TagOut, Write, Clk);

input    ['INDEX]Address;
input    ['TAG]    TagIn;
output   ['TAG]    TagOut;
input              Write;
input              Clk;

reg      ['TAG]    TagOut;
reg      ['TAG]    TagRam    [0:'CACHESIZE-1];

always @ (negedge Clk)
      if (Write)
      TagRam[Address]=TagIn; // write

always @ (posedge Clk)
            TagOut = TagRam[Address]; // read

endmodule
```

**Figure 5.11** Tag RAM Module

**Valid RAM Model**

The ValidRAM is a memory array containing a single bit for each index location indicating whether that location contains valid data. Figure 5.12 shows how each bit in ValidBits corresponds to an entry in the Tag RAM. The main difference between the ValidRAM model and the TagRAM model is that all bits in ValidBits need to be reset to zero when Reset is asserted.

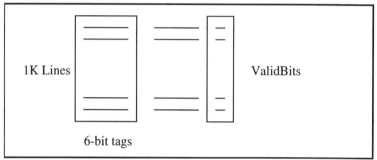

**Figure 5.12** Tag RAM and ValidBits

Figure 5.13 contains the code used to implement the RAM for the ValidBits. Note that this module is not synthesizable.

```
module ValidRam(Address, ValidIn, ValidOut,
      Write, Reset, Clk);
input   ['INDEX]    Address;
input               ValidIn;
output              ValidOut;
input               Write;
input               Reset;
input               Clk;
reg                 ValidOut;
reg     ['CACHESIZE-1:0] ValidBits;
integer i;
always @ (posedge Clk) // Write
      if (Write && !Reset)
            ValidBits[Address]=ValidIn; // write
      else if (Reset)
            for (i=0;i<'CACHESIZE;i=i+1)
                  ValidBits[i]='ABSENT; // reset
always @ (posedge Clk) // REad
            ValidOut = ValidBits[Address]; // read

endmodule
```

**Figure 5.13** RAM for ValidBits

# Modeling Cache Memories

**Data RAM Model**

The Data RAM model is simply a memory array containing eight bits of data for each line in the cache. The Data RAM is identical in function and timing to the Tag RAM but it is 16 bits wide as compared to 6 bits wide in the Tag RAM. The code for the Data RAM model is shown in Figure 5.14. Note that this module is not synthesizable.

```
module DataRam(Address, DataIn,
                DataOut, Write, Clk);
input    ['INDEX]Address;
input    ['DATA]    DataIn;
output   ['DATA]    DataOut;
input               Write;
input               Clk;
reg      ['DATA]    DataOut;
reg      ['DATA] DataRam ['CACHESIZE-1:0];
always @ (posedge Clk)
        if (Write)
                DataRam[Address]=DataIn; // write
always @ (posedge Clk)
                DataOut = DataRam[Address]; // read

endmodule
```

**Figure 5.14** Code for Data RAM

**Tag Comparator**

The Tag Comparator compares the output of the TagRam to the Tag portion of the processor address. The resulting output signal is used to determine if the data for the requested address is actually stored in that line of the cache. Figure 5.15 shows the code for the Tag Comparator. Note that "==" is used rather than "===" so that unknown values are propagated to the Match output correctly.

```
module Comparator(Tag1,Tag2,Match);
        input    ['TAG]    Tag1;
        input    ['TAG]    Tag2;
        output             Match;
        wire     Match = Tag1 == Tag2;

endmodule
```

**Figure 5.15** Tag Comparator Module

# Digital Design and Synthesis with Verilog HDL

**Data Multiplexors**

Data to be written into the cache data RAMs can come from either the processor or from the system bus. Data that is returned to the processor may come from either the cache or from the system bus. Two multiplexors are used, each one having the full width of the data bus. One of the data multiplexors is used to select the source of the input data to the Data RAM, the other is used to select the source of the data to be returned to the processor. Figure 5.16 shows the code for the data multiplexors. The multiplexor is defined using a continuous assignment statement.

```
module DataMux(S,A,B,Z);

    input               S;  // Select line
    input   ['DATA]     A;  // A input bus
    input   ['DATA]     B;  // B Input bus
    output  ['DATA]     Z;  // output bus

    wire    ['DATA]     Z = S ? A : B;
endmodule
```

**Figure 5.16** Data Multiplexors

**Controller Model**

The controller consists of a state machine and a small counter. The state diagram for the controller is shown in Figure 5.17. The state machine coordinates all of the activities within the chip, for example, controlling system bus transactions, asserting PReady and tri-stating the data buses at the appropriate time. The counter is used for counting cycles for wait states that are inserted into transactions on the system bus.

During IDLE the state machine waits for the processor to start a cycle with the assertion of PStrobe. This signal is sampled at the rising edge of the clock. When PStrobe is asserted, the address is looked up in the cache. If the tag that is stored in the Tag RAM for this address matches the tag of the current processor address and the corresponding ValidBit is set, then data for this address is stored in the cache. Based on whether the data is in the cache and on the state of the PRW line, the correct sequence of subsequent actions are determined. The four possibilities and their associated actions are

# Modeling Cache Memories

1. Read Hit, return data from cache.
2. Read Miss, fetch data from memory, return data and update cache.
3. Write Hit, write to cache and to memory.
4. Write Miss, write to memory only.

Each of these cases is described in detail later in this chapter.

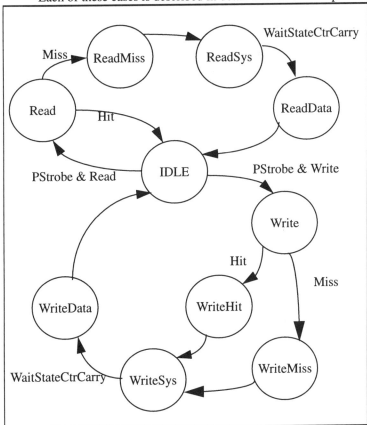

**Figure 5.17** State diagram of controller

Figure 5.17 shows a state diagram for the control state machine. Note that all transactions on the system bus take a set number of cycles regardless of the type of transaction. During each of the bus operations, the wait state counter is loaded. The state machine does not advance state

# Digital Design and Synthesis with Verilog HDL

until the counter has overflowed. The number of wait states, i.e., additional cycles, that the bus needs is defined by the identifier WAITSTATES. At the end of each bus transaction, the controller asserts PReady for one cycle to indicate to the processor that the transaction is finished.

**Read Hit**

Of all the possible cache operations, read hits happen most often because of the locality of bus references for a typical software program and because reads are more common than writes. In the cycle following the processor strobe, data is returned to the processor from the Data RAM along with the assertion of PReady. Since the state machine is unable to generate PReady without the result of the TagRam reference, some external gating is necessary. The state machine generates a signal PReadyEnable that is gated with the Hit signal to generate PReady. In some of the other cases PReady needs to be generated directly from the state machine. Figure 5.18 shows a schematic of how this is acheived.

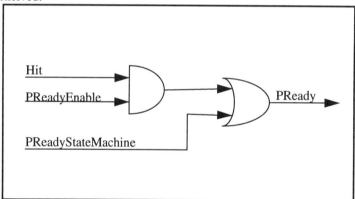

**Figure 5.18** PReady generation

**Write Hit**

Since the cache protocol is write-through, both the cache and the main memory must be updated. Main memory is updated at the same time to maintain cache coherency. This simply means that main memory always has a copy of the most recent data. A write cycle must be performed on the system bus at the same time that data is written to the cache. After the preset number of wait states, PReady is returned to the processor to indicate the completion of the transaction. The Tag RAM does not need to be updated or modified.

**Read Miss**

A read miss occurs if the location in the cache corresponding to the current processor address is either invalid or contains data for a different address. A memory cycle reads data from the main memory. After a set number of wait states have passed, the data is available for writing into the cache and for returning to the processor. At the same time, the Tag RAM is updated to reflect the tag of the new data and the ValidBit is set.

**Write Miss**

In the case of a write miss, only the main memory is updated. This policy is chosen because the data read is more likely to be used again than data that is written. The model performs a write to memory only and returns `PReady` when the bus transaction is complete.

**State Machine**

There are two different ways to approach the implementation of the state machine. One is to write the state machine directly in Verilog HDL and the other is to use a state machine compiler that might be provided with a synthesis tool. We will examine the Verilog HDL approach because this provides the easiest method for simulation.

Figure 5.19 shows the Verilog code to implement the state machine.

```
module Control(
    PStrobe, PRW, PReady,
    Match, Valid,
    Write,
    CacheDataSelect,
    PDataSelect,
    SysDataOE, PDataOE,
    SysStrobe, SysRW, Reset, Ph1, Ph2
    );                                  continued
```

**Figure 5.19** Code For Controller State Machine

```
        input       PStrobe, PRW;
        output      PReady;
        input       Match, Valid;
        output      Write;
        output      CacheDataSelect;
        output      PDataSelect;
        output      SysDataOE, PDataOE;
        output      SysStrobe, SysRW;
        input       Reset;
        input       Clk;
        wire  [1:0] WaitStateCtrInput = 'WAITSTATES -1;
        reg         LoadWaitStateCtr;

        WaitStateCtr WaitStateCtr(
                .Load       (LoadWaitStateCtr),
                .LoadValue  (WaitStateCtrInput),
                .Carry      (WaitStateCtrCarry),
                .Ph1        (Ph1),
                .Ph2        (Ph2)
            );

        reg         PReadyEnable;
        reg         SysStrobe;
        reg         SysRW;
        reg         SysDataOE;
        reg         Write;
        reg         Ready;
        reg         CacheDataSelect;
        reg         PDataSelect;
        reg         PDataOE;

        reg   [3:0] State;
        reg   [3:0] NextState;
        initial State = 0;
        parameter STATE_IDLE = 0,
              STATE_READ = 1,
              STATE_READMISS = 2,
              STATE_READSYS = 3,
              STATE_READDATA = 4,
              STATE_WRITE = 5,
              STATE_WRITEHIT = 6,
              STATE_WRITEMISS = 7,
              STATE_WRITESYS = 8,
              STATE_WRITEDATA = 9;
```

**continued**

**Figure 5.20** Code For Controller State Machine (continued)

```verilog
always @ (posedge Clk) State = NextState;
always @ (State)
if (Reset) NextState = 'STATE_IDLE;
else
case (State)//synopsys parallel_case full_case
     'STATE_IDLE: begin
              if (PStrobe && PRW)
                    NextState = 'STATE_READ;
              else if (PStrobe && !PRW)
                    NextState = 'STATE_WRITE;
end
'STATE_READ : begin
              if (Match && Valid)
                    NextState = 'STATE_IDLE;
              // read hit
              else
                    NextState = 'STATE_READMISS;
              $display ("state = read");
end
'STATE_READMISS : begin
              NextState = 'STATE_READSYS;
              $display ("state = readmiss");
end

'STATE_READSYS : begin
              if (WaitStateCtrCarry)
                    NextState = 'STATE_READDATA;
              else
                    NextState = 'STATE_READSYS;
              $display ("state = readsys");
end

'STATE_READDATA : begin
              NextState = 'STATE_IDLE;
              $display ("state = readdata");
end

'STATE_WRITE : begin
              if (Match && Valid)
                    NextState = 'STATE_WRITEHIT;
              else
                    NextState = 'STATE_WRITEMISS;
              $display ("state = WRITE");
end                                              continued
```

**Figure 5.21** Code For Controller State Machine (continued)

```verilog
            'STATE_WRITEHIT : begin
                    NextState = 'STATE_WRITESYS;
                    $display ("state = WRITEHIT");
            end

            'STATE_WRITEMISS : begin
                    NextState = 'STATE_WRITESYS;
                    $display ("state = WRITEmiss");
            end

            'STATE_WRITESYS : begin
                    if (WaitStateCtrCarry)
                            NextState = 'STATE_WRITEDATA;
                    else
                            NextState = 'STATE_WRITESYS;
                    $display ("state = WRITEsys");
            end

            'STATE_WRITEDATA : begin
                    NextState = 'STATE_IDLE;
                    $display ("state = WRITEdata");
            end
endcase
task OutputVec;
input [9:0] vector;
begin
        LoadWaitStateCtr=vector[9];
        PReadyEnable=vector[8];
        Ready=vector[7]
        Write=vector[6];
                SysStrobe[5];
        SysRW=vector[4];
        CacheDataSelect=vector[3];
        PDataSelect=vector[2];
        PDataOE=vector[1];
        SysDataOE=vector[0];
end
endtask
always @ (State)
case (State)
'STATE_IDLE:          OutputVec(10'b0000000000);
'STATE_READ :         OutputVec(10'b0100000010);
'STATE_READMISS :     OutputVec(10'b1000110010);
'STATE_READSYS :      OutputVec(10'b0000010010);
'STATE_READDATA :     OutputVec(10'b0011011110);
                                              continued
```

**Figure 5.22** Code For Controller State Machine (continued)

```
'STATE_WRITEHIT :    OutputVec(10'b1001101100);
'STATE_WRITE :       OutputVec(10'b0100000000);
'STATE_WRITEMISS :   OutputVec(10'b1000100001);
'STATE_WRITESYS :    OutputVec(10'b0000010001);
'STATE_WRITEDATA :   OutputVec(10'b0011011101);
endcase

wire PReady =(PReadyEnable && Match && Valid)
       ||    Ready;

endmodule
```

**Figure 5.23** Code for Controller State Machine (continued)

## Wait State Counter

The `WaitStateCtr` counter is a 2-bit down counter that generates a carry output signal when the count reaches zero. The code for the wait state counter is shown in Figure 5.24.

```
module WaitStateCtr(Load, LoadValue, Carry, Clk);
input           Load;
input   [1:0]   LoadValue;
output          Carry;
input           Clk;

reg     [1:0]   Count;

always @ (posedge Clk)
    if (Load)
        Count = LoadValue;
    else
        Count = Count - 1;

wire Carry = Count == 2'b0;

endmodule
```

**Figure 5.24** Wait State Counter

# Digital Design and Synthesis with Verilog HDL

## Testing

Testing the cache, as a stand-alone unit, requires writing a special driver to simulate processor operation. The code for the cache driver instantiates one copy of the cache model and includes tasks for read and write operations as if they had come from the processor. The code includes a simple model for main memory as well. At the end of the module, a short test first writes data to a location, then reads that location several times. The write operation writes through to the memory without updating the cache, and the first read operation loads the cache with data for that address. The second read operation is a cache hit and is thus much faster than the first read operation.

The Verilog command to run this test is

```
> verilog cache.v cachedriver.v
```

Figure 5.25 shows the waveforms generated by this short test. The waveforms obtained from running the simulation demonstrate that the cache model works correctly. More stringent tests should be written to fully prove the functionality of the model. Given Verilog tasks for reading and writing data, the task of fully testing the cache is reduced to a series of loops to check writing and reading all memory locations. If trace data, i.e., address traces from running real programs on the processor were available, this model could be used for performance evaluations for the cache size and architecture chosen.

## Performance Improvements

All real cache systems have the mechanisms demonstrated in the foregoing simple model. Some suggestions are offered here for improving the performance of the basic model for real realistic applications.

### Two-Way Set Associative Cache

The simplest way to describe two-way set associativity is to say that it is like having two direct-mapped caches similar to the one in the model. Both Tag RAMs are looked up simultaneously, and data is provided from the cache in which the address generates a hit. The data cannot reside in both caches because it is never written into both caches. During a replacement, one of the caches receives the data, chosen either randomly or by a least-recently-used algorithm.

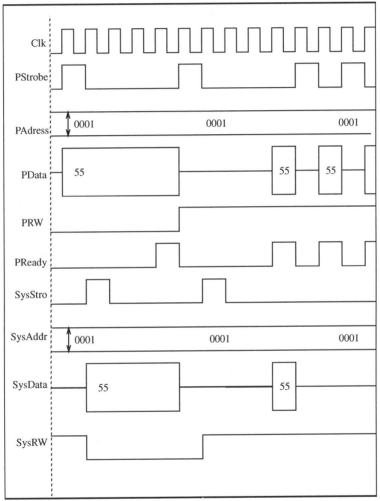

**Figure 5.25** Result of short test

Multi-way set associative caches relieve a problem called "thrashing". Thrashing occurs when a program accesses two different memory locations that map to the same location in the cache. A two-way set associative cache provides two places where one address can be mapped. The model remains almost the same, but the Tag and Data RAMs are duplicated and a tag replacement strategy is implemented.

Higher associativities are simply extensions of this same principle and provide further improvements in the performance. The actual improvement depends on the type of code and the cache size.

**Write Buffering**

Write buffering is, in a way, a mechanism that enables the cache subsystem to "lie" to the processor about the completion of write operations. As the cache controller starts a write on the system bus it returns Ready immediately to the processor before the write completes. The processor then proceeds and, provided that it does not do a transaction that requires the system bus before the current write is completed, it gains valuable cycles. For example, if a write is followed by a read that hits in the cache, the read takes place while the write onto the system bus is still being completed.

To upgrade the model to allow for write buffering, the code for the write must be modified so that PReady is returned immediately after the system bus transaction begins. Note that another system transaction cannot be started until the current one is completed.

**Larger Line Size**

The line size is the number of words stored in the cache for each tag. In this chapter's simple example the line size is 1. In higher performance systems the line size can be much larger. Sometimes the system bus is capable of transferring the entire line during a single bus transaction. In this case the number of words that the system bus can supply in one transaction is called the "block size". A larger block size improves bus performance by reducing the average overhead for bus accesses. Implementing a larger line size in our model simply means reducing the size of the Tag RAM. For example, if the line size were 4 words, a 1K cache would require only 256 tags. The Tag RAM would be indexed by PAddress[7:0]. To compensate for the reduced number of tags, each tag would then be 8 bits corresponding to PAddress[16:8]. As can be seen, a larger block size could be implemented with a few trivial changes to the basic model.

If the bus is not capable of multiword transfers, a larger block size can still offer an advantage. During a cache miss, either the full block must be replaced one word at a time, or only the word that was requested is obtained and a valid bit is set. A ValidBit is used for each word in the block indicating which words of the block are currently present in the cache. Larger block sizes in this case reduce the number of tags stored in the Tag RAM.

**Write-Back Policy**

Instead of always writing to memory, it is possible to write data back to the memory only when a replacement would write over that location in the cache. An extra bit is required, along with the valid bit, for an indication of whether each location in the cache has been written to. Main memory is no longer coherent with the cache. Write-back caches are more complicated than write-through caches, but they provide better performance in multi-processor systems. Complexities arise in multi processor systems because of the difficulty in keeping multiple caches coherent with each other.

To upgrade the basic cache model to a write-back cache in our single processor system requires only the addition of the extra bit mentioned above to indicate modified lines. The code for the write hit case would change to just write into the cache and set the bit that indicates modification. This bit can be referred to as the "dirty bit". The data would not be written through to the system bus until a replacement needed the same cache location.

Write-back caches are generally used in multi-processing environments. The implementation of such a cache system is beyond the scope of the basic model and would require a number of additional signals on the system bus for maintaining coherency between multiple caches on the same bus.

# Summary

In this chapter a model of a direct-mapped cache was presented to demonstrate some of the issues involved with writing Verilog for use with synthesis tools. Variations of this model can be used to evaluate the performance of different cache architectures and with little effort be

Digital Design and Synthesis with Verilog HDL

converted to a gate level netlist by synthesizing into a specific technology or vendor library.

## Exercises

1. A framework for testing has been provided although the testing implemented is somewhat minimal. Improve the testing for the cache by adding functions to write and read all addressable locations. We suggest that you make two passes. One with the data having the same bit pattern as the address and the other with the data equal to the bit-wise inversion of the address.

2. In a real cache system it can be advantageous to provide a mechanism for directly testing the TagRAM. This can be accomplished in many ways. The simplest is to provide an extra input with the same logical timing as PRW to indicate that the access should be directed to the TagRAM. This signal could be derived from decoding some of the high order address bits, for example. Add the code necessary to transfer PData to and from the addressed TagRAM line.

3. This model can be used for testing the performance of the Cache system. Add the code necessary to generate the Cache statistics, i.e. the Cache hit to miss ratio, total number of cycles, etc. You can generate addresses randomly or use an address trace from a real program. By varying the number of wait states on the system bus you will be able to see the effect of main memory latency. You can modify the size of the Cache and rerun your traces to determine the optimal Cache size.

4. For a limited TagRAM size, performance can be improved by increasing the line size. Try increasing the line size of the model to 4. You may need to add extra ValidBits to indicate which of the words within a line are valid.

5. Investigate the performance improvement gained by adding a write buffer. Naturally the improvement will depend on the mix of reads and writes in your address trace. The Performance Improvements section describes a one level write buffer. Write buffers can be of any depth. Add to the model a variable depth write buffer and rerun your traces to determine the optimal write buffer size.

6. Investigate performance gains from using a write-back policy. You will need to add an extra bit to the TagRAM to indicate that the line is dirty and will need to be written back to main memory before it can be replaced.

7. Further improvements to performance can be attained by changing the model to a multi-way set associative cache. Try increasing the model to a four way set associative model. You will need to adopt an appropriate algorithm for deciding which of the lines will be replaced in the event that they have all been filled. Random selection is by far the simplest.

8. Many synthesis programs provide state machine tools for determining the ideal assignment of the states. If you have such a program available, try to find the best assignment of states for the control state machine.

```verilog
// Defines

`define READ 1'b1
`define WRITE 1'b0
`define CACHESIZE 1024
`define WAITSTATES 2'd2
`define ADDR 15:0
`define ADDRWIDTH 16
`define INDEX 9:0
`define TAG 15:10
`define DATA 31:0
`define DATAWIDTH 32
`define PRESENT 1'b1
`define ABSENT !`PRESENT

module cache(
        PStrobe,
        PAddress,
        PData,
        PRW,
        PReady,

        SysStrobe,
        SysAddress,
        SysData,
        SysRW,

        Reset,
        Clk
        );

input            PStrobe;
input   [`ADDR]  PAddress;
inout   [`DATA]  PData;
input            PRW;
output           PReady;

output           SysStrobe;
output  [`ADDR]  SysAddress;
inout   [`DATA]  SysData;
output           SysRW;
input            Reset;
input            Clk;
```

**continued**

**Figure 5.26** Model of the cache

```verilog
// Bidirectional Buses
wire    PDataOE;
wire    SysDataOE;
wire['DATA]PDataOut;
wire    ['DATA] PData=
        PDataOE ? PDataOut : 'DATAWIDTH'bz;
wire    ['DATA] SysData=
        SysDataOE ? PData : 'DATAWIDTH'bz;

wire['ADDR] SysAddress = PAddress;
wire['TAG]TagRamTag;

wire    Write;
wire    Valid;
wire    CacheDataSelect;
wire    PDataSelect;
wire    Match;

TagRam TagRam(
        .Address(PAddress['INDEX]),
        .TagIn      (PAddress['TAG]),
        .TagOut     (TagRamTag['TAG]),
        .Write      (Write),
        .Clk        (Clk)
    );

ValidRam ValidRam(
        .Address(PAddress['INDEX]),
        .ValidIn(1'b1),
        .ValidOut(Valid),
        .Write      (Write),
        .Reset      (Reset),
        .Clk        (Clk)
    );

wire['DATA]DataRamDataOut;
wire['DATA]DataRamDataIn;

DataMux CacheDataInputMux(
        .S      (CacheDataSelect),
        .A      (SysData),
        .B      (PData),
        .Z      (DataRamDataIn)
    );
```
**continued**

**Figure 5.27** Model of the cache (continue)

```
DataMux PDataMux(
        .S      (PDataSelect),
        .A      (SysData),
        .B      (DataRamDataOut),
        .Z      (PDataOut)
  );

DataRam DataRam(
        .Address(PAddress['INDEX]),
        .DataIn    (DataRamDataIn),
        .DataOut(DataRamDataOut),
        .Write     (Write),
        .Clk       (Clk)
  );

Comparator Comparator(
        .Tag1   (PAddress['TAG]),
        .Tag2   (TagRamTag),
        .Match  (Match)
  );

Control Control(
        .PStrobe(PStrobe),
        .PRW       (PRW),
        .PReady    (PReady),
        .Match     (Match),
        .Valid     (Valid),
        .CacheDataSelect (CacheDataSelect),
        .PDataSelect(PDataSelect),
        .SysDataOE(SysDataOE),
        .Write     (Write),
        .PDataOE(PDataOE),
        .SysStrobe(SysStrobe),
        .SysRW     (SysRW),
        .Reset     (Reset),
        .Clk       (Clk)
  );

endmodule
```

**continued**

**Figure 5.28** )Model of the cache (continued)

# Modeling Cache Memories

```
module Control(
PStrobe, PRW, PReady,
Match, Valid,
Write,
CacheDataSelect,
PDataSelect,
SysDataOE, PDataOE,
SysStrobe, SysRW, Reset,
Clk
);

input PStrobe, PRW;
output PReady;
input Match, Valid;
output Write;
output CacheDataSelect;
output PDataSelect;
output SysDataOE, PDataOE;
output SysStrobe, SysRW;
input Reset;
input Clk;

wire [1:0]WaitStateCtrInput = 'WAITSTATES - 2'd1;
wire    WaitStateCtrCarry;
reg     LoadWaitStateCtr;

WaitStateCtr WaitStateCtr(
.Load      (LoadWaitStateCtr),
.LoadValue (WaitStateCtrInput),
.Carry     (WaitStateCtrCarry),
.Clk       (Clk)
);

reg PReadyEnable;
reg SysStrobe;
reg SysRW;
reg SysDataOE;
reg Write;
reg Ready;
reg CacheDataSelect;
reg PDataSelect;
reg PDataOE;
reg [3:0]State;
```
**continued**

**Figure 5.29** Model of the cache (continued)

```
reg[3:0]NextState;

'define STATE_IDLE 4'd0
'define STATE_READ 4'd1
'define STATE_READMISS 4'd2
'define STATE_READSYS 4'd3
'define STATE_READDATA 4'd4
'define STATE_WRITE 4'd5
'define STATE_WRITEHIT 4'd6
'define STATE_WRITEMISS 4'd7
'define STATE_WRITESYS 4'd8
'define STATE_WRITEDATA 4'd9

always @(posedge Clk)
State = Reset ? 'STATE_IDLE : NextState;

always @ (State or
        PStrobe or
       PRW or
       Match or
       Valid or
       WaitStateCtrCarry)
       case (State)
       'STATE_IDLE: begin
            if (PStrobe && PRW=='READ)
                NextState = 'STATE_READ;
            else if (PStrobe && PRW=='WRITE)
                NextState = 'STATE_WRITE;
            else
                NextState = 'STATE_IDLE;
       end
       'STATE_READ : begin
            if (Match && Valid)
                NextState = 'STATE_IDLE;
                            // read hit
            else
                NextState = 'STATE_READMISS;
       end
       'STATE_READMISS : begin
            NextState = 'STATE_READSYS;
       end
       'STATE_READSYS : begin
            if (WaitStateCtrCarry)
                NextState = 'STATE_READDATA;
```

**continued**

**Figure 5.30** Model of the cache (continued)

```verilog
                    else
                            NextState = 'STATE_READSYS;
                end
                'STATE_READDATA : begin
                    NextState = 'STATE_IDLE;
                end
                'STATE_WRITE : begin
                    if (Match && Valid)
                            NextState = 'STATE_WRITEHIT;
                    else
                            NextState = 'STATE_WRITEMISS;
                end
                'STATE_WRITEHIT : begin
                    NextState = 'STATE_WRITESYS;
                end
                'STATE_WRITEMISS : begin
                    NextState = 'STATE_WRITESYS;
                end
                'STATE_WRITESYS : begin
                    if (WaitStateCtrCarry)
                            NextState = 'STATE_WRITEDATA;
                    else
                            NextState = 'STATE_WRITESYS;
                end
                'STATE_WRITEDATA : begin
                    NextState = 'STATE_IDLE;
                end
                default:
                    NextState = 'STATE_IDLE;
                endcase

task OutputVec;
input [9:0] vector;
begin
{LoadWaitStateCtr, PReadyEnable, Ready,
 Write, SysStrobe, SysRW, CacheDataSelect,
        PDataSelect, PDataOE, SysDataOE} = vector;
end
endtask

always @ (State)
case (State)
        'STATE_IDLE: OutputVec(10'b0000000000);
```

**continued**

**Figure 5.31** Model of the cache (continued)

```
                'STATE_READ : OutputVec(10'b0100000010);
                'STATE_READMISS : OutputVec(10'b1000110010);
                'STATE_READSYS : OutputVec(10'b0000010010);
                'STATE_READDATA : OutputVec(10'b0011011110);
                'STATE_WRITEHIT : OutputVec(10'b1001101100);
                'STATE_WRITE : OutputVec(10'b0100000000);
                'STATE_WRITEMISS : OutputVec(10'b1000100001);
                'STATE_WRITESYS : OutputVec(10'b0000000001);
                'STATE_WRITEDATA : OutputVec(10'b0011001101);
                default :   OutputVec(10'b0000000000);
    endcase

wire PReady = (PReadyEnable && Match && Valid) ||
                                               Ready;

endmodule

module DataMux(S, A, B, Z);

input S; // Select line
input ['DATA] A;
input ['DATA] B;
output ['DATA] Z;

wire ['DATA] Z = S ? A : B;

endmodule

module WaitStateCtr(Load, LoadValue, Carry, Clk);

input Load;
input[1:0]LoadValue;
outputCarry;
inputClk;
reg[1:0] Count;

always @(posedge Clk)
  if (Load)
        Count = LoadValue;
  else
        Count = Count - 2'b1;

wire Carry = Count == 2'b0;
endmodule
```

**continued**

**Figure 5.32** Model of the cache (continued)

```
// Macros

module TagRam(Address, TagIn, TagOut, Write, Clk);

input ['INDEX]Address;
input['TAG]TagIn;
output['TAG]TagOut;
input    Write;
input    Clk;

reg['TAG]TagOut;
reg      ['TAG]  TagRam  ['CACHESIZE-1:0];

always @(negedge Clk)
  if (Write)
       TagRam[Address]=TagIn; // write
  else
       ;

always @(posedge Clk)
  TagOut = TagRam[Address]; // read

endmodule

module ValidRam(Address, ValidIn, ValidOut, Write,
                                      Reset, Clk);

input ['INDEX]Address;
input    ValidIn;
output   ValidOut;
input    Write;
input    Reset;
input    Clk;

reg      ValidOut;
reg      ['CACHESIZE-1:0] ValidBits;

integer i;

always @ (posedge Clk)
  if (Write && !Reset)
       ValidBits[Address]=ValidIn; // write
```

**continued**

**Figure 5.33** Model of the cache (continued)

```
       else if (Reset)
           for (i=0;i<'CACHESIZE;i=i+1)
               ValidBits[i]='ABSENT; // reset
     else
           ;

  always @ (posedge Clk)
    ValidOut = ValidBits[Address]; // read

  endmodule

  module DataRam(Address, DataIn, DataOut, Write, Clk);

  input ['INDEX]Address;
  input['DATA]DataIn;
  output['DATA]DataOut;
  input    Write;
  input    Clk;

  reg['DATA]DataOut;
  reg      ['DATA] Ram ['CACHESIZE-1:0];

  always @ (posedge Clk)
    if (Write)
         Ram[Address]=DataIn; // write
    else
         ;

  always @ (posedge Clk)
    DataOut = Ram[Address]; // read

  endmodule

  module Comparator(Tag1, Tag2, Match);

  input['TAG]Tag1;
  input['TAG]Tag2;
  output   Match;

  wireMatch = Tag1 == Tag2;

  endmodule
```

**Figure 5.34** Model of the cache (continued)

CHAPTER

# 6

# Modeling Asynchronous I/O: UART

In this chapter we present an example of modeling an asynchronous peripheral device, a dual Universal Asynchronous Receiver Transmitter (UART) chip. We develop two models of the chip. The first model is a high-level abstraction which describes the functionality of the chip and emphasizes simplicity, readability, and ease of change. The second model is oriented toward gate-level implementation. This model is partitioned so that a logic synthesizer can be used to automatically implement the chip with library components.

**Functional Description of the UART**

A UART is used for communicating with serial input/output devices. Serial communication is needed either when the device is inherently serial (e.g. modems and telephone lines) or when the cabling cost has to be reduced at the expense of operating speed (e.g., a twisted pair in laboratory instrumentation).

Typically, the UART is connected between a central processor and a serial device. To the processor, the UART appears as an 8-bit parallel port which can be read from or written to. To the serial device, the UART presents two data wires, one for input and one for output, which serially communicate 8-bit data. The rate of data communication depends on the peripheral device. Some devices operate at a single clock speed (e.g., old teletypes at 110 baud), and they generate the clock internally. Other devices can operate at multiple clock rates and get clock input from the UART.

To detect the start of a transmitted byte on the serial line, the line is held high between successive transmissions and is pulled low for a one-bit duration. Other issues relating to serial communications—such as parity, overruns, and stop bits—are not covered in this example.

Figure 6.2 shows a block diagram of the dual UART chip. It has two identical single UART modules which operate independently of each other. All the input/output pins of the chip are listed in Figure 6.1.

```
Signal name      Signal description        I/O
----------------------------------------------
reset            reset                     I
clkin            external clock            I
rd_              read signal               I
wr_              write signal              I
cs_0, cs_1       chip selects              I
din0, din1       serial data input         I
a[2:0]           address bus               I
dbus[7:0]        bidirectional data bus    I/O
int0, int1       interrupt lines           O
dout0, dout1     serial data outputs       O
clko0, clko1     output clocks             O
```

**Figure 6.1**  I/O pins of the UART chip

Each UART module has eight 8-bit registers for control and status. The registers can be read from and written to using the `rd_` and `wr_` signals. The a input is an address to select one of the eight registers to read or write. To improve the readability of the model, the register addresses have been assigned mnemonics. For example, address 0, which is the location of the transmit data register, is designated as `XMITDT_ADDR`. Figure 6.3 shows the mnemonics of the various addresses.

Waveforms for reading from and writing to an internal register are shown in Figure 6.4 and Figure 6.5.

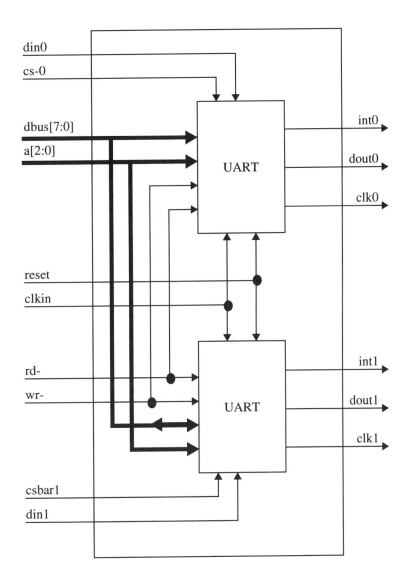

**Figure 6.2** The dual UART Chip

```
        // The addresses of the internal registers
        parameter
                XMITDT_ADDR = 0,
                STATUS_ADDR = 1,
                DIVLSB_ADDR = 2,
                DIVMSB_ADDR = 3,
                RECVDT_ADDR = 4,
                CLRINT_ADDR = 7;
```

**Figure 6.3** Register address mnemonics

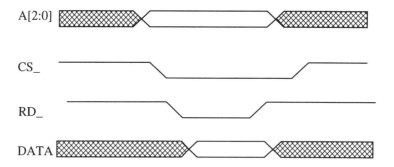

**Figure 6.4** Reading from an internal register

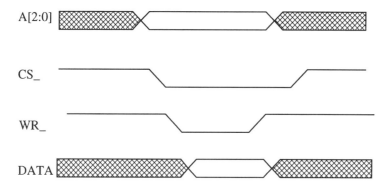

**Figure 6.5** Writing to an internal register

## Modeling Asynchronous I/O: UART

Serial transmission is initiated when the host processor writes to register XMITDT of the UART. Reception is triggered by a falling edge of the serial data line (din). While transmitting or receiving, register STATUS reflects the current status. When either the transmission or the reception is complete, an interrupt is generated. The interrupt is cleared when register CLRINT of the corresponding UART is read. The status indicates the source of the interrupt (receive or transmit).

Registers DIVMSB and DIVLSB contain the most- and least-significant bits of a 16-bit integer, which divides the external clock to get the desired baud rate. Registers RECVDT and XMITDT contain the receive and transmit data. The rest of the registers in the register file are not used in this model; however, in a more complex design, they can hold other parameters such as parity, number of stop bits, overflow, and so forth.

Figure 6.6 shows the top-level structural model of the dual UART chip. It instantiates two single UART modules and follows the block diagram closely.

```
module dual_uart (
    dbus, a, reset, rd_, wr_,
    cs_0, cs_1, din0, din1, clkin,
    int0, int1, dout0, dout1, clko0, clko1
);
inout[7:0] dbus;
input[2:0] a;
input reset, rd_, wr_;
input cs_0, cs_1, din0, din1, clkin;

output int0, int1, dout0, dout1, clko0, clko1;

// The first instance of a single UART module
uart u0 (
    reset, dbus, a, rd_, wr_, cs_0, din0, clkin,
    int0, dout0, clko0
);

// The second instance of a single UART module
uart u1 (
    reset, dbus, a, rd_, wr_, cs_1, din1, clkin,
    int1, dout1, clko1
);
endmodule
```

**Figure 6.6** Structural model of the dual UART

The next few sections describe the model of the single UART module.

## Functional Model of the Single UART

In the functional model, we are concerned with the ease of modeling, readability, and maintainability. We disregard any implementation details. We use an array of registers to describe the internal status and control registers of the chip, which are convenient to program but might not be efficient to implement. We also use events and triggers, Verilog constructs which are difficult to synthesize. The functional model is shown in Figure 6.8.

```
always @(negedge reset) begin : reset_block
    integer i;

    disable receive_block;
    disable transmit_block;
    oclkreg = 0;
    iclkreg = 0;
    int = 0;
    dbus_reg = 8'hzz;
    for (i = 0; i < 8; i = i + 1)
        regfile[i] = 0;
end
```

**Figure 6.7** Reset operation

In the next sections we describe the main blocks in the model: the reset block, the clock generator, the read block, the write block, the transmit block and the receive block.

**Reset Operation**

Reset is initiated when the `reset` signal makes a negative transition. On `reset`, the model disables the receive and transmit blocks, initializes the clocks, sets all the registers in the register file to 0, and tristates the outputs. The code is shown in Figure 6.7.

**Clock Generator**

The UART generates both the transmit clock and the receive clock. The two clocks have the same frequency but may be shifted in phase with respect to each other. The external clock is divided by the

```
module uart (
     reset, dbus, a, rd_, wr_, cs_, din, clkin,
     int, dout, clko
);
// Input/output declarations
...

// The addresses of the internal registers
parameter ...

// The internal state
parameter
   TRANSMITTING = 0,
   RECEIVING = 1,
   DONE_XMT = 2,
   DONE_RCV = 3;

reg[7:0] regfile [0:7];

wire[15:0] divisor =
  { regfile[DIVMSB_ADDR], regfile[DIVLSB_ADDR] };

reg[15:0] oclkreg, iclkreg;

reg xmt_clock, rcv_clock;

event do_transmit;

integer lines;


// All the functional blocks of the module
// are described in the following subsection

// Reset operation
...
// Clock generator
...
// Read operation
...
// Write operation
...
// Transmit operation
...
// Receive operation
...
endmodule // single_uart
```

**Figure 6.8** Functional model of the single UART

divisor, which is the concatenation of DIVMSB and DIVLSB. The receive and transmit clocks are generated in parallel, using the fork-join construct of Verilog as shown in Figure 6.9. Fork-join is not synthesizable in existing synthesis tools.

```
always @(posedge clkin) begin
    oclkreg = oclkreg + 1;
    iclkreg = iclkreg + 1;
    fork
    if (oclkreg >= divisor) begin
       oclkreg = 0;
       xmt_clock = 1;
       clko = 1;
       #1
       clko = 0;
       xmt_clock = 0;
    end
    if (iclkreg >= divisor) begin
       iclkreg = 0;
       rcv_clock = 1;
       #1
       rcv_clock = 0;
    end
    join
end
```

**Figure 6.9** Clock generator

**Read Operation**

The read operation reads an internal register from the register file of the UART. Reading occurs on the negative edge of rd_. At that time, the model writes to the appropriate register on the bus and, on the positive edge of rd_, it tristates the bus. In addition, if the address being read is the clear-interrupt address (CLRINT_ADDR), the model resets the DONE_XMT and DONE_RCV bits of the status and clears the int output. This operation is shown in Figure 6.10.

**Write Operation**

Writing into a register is initiated on the negative edge of the wr_ input. The model loads the appropriate register with the value from the data bus. If the address is XMITDT_ADDR, the do_transmit event is triggered and a transmission of a byte is initiated. This operation is shown in Figure 6.11.

# Modeling Asynchronous I/O: UART

```
always @(negedge rd_) if (~cs_) begin
    dbus_reg = regfile[a];
    @(posedge rd_)
    dbus_reg = 8'hzz;
    if (a == CLRINT_ADDR) begin : read_block
        int = 0;
        // Reset the DONE_XMT and DONE_RCV bits in
        // the status register
        ...
    end
end
```

**Figure 6.10** Read operation

```
always @(negedge wr_) if (~cs_) begin
    @(posedge wr_)
    regfile[a] = dbus;
    if (a == XMITDT_ADDR) ->do_transmit;
end
```

**Figure 6.11** Write operation

**Transmit Operation**

Transmission is initiated by triggering the do_transmit event. First, the transmit clock is initialized, and the eight bits of the byte are transmitted serially. Second, the model sets the done bit and generates an interrupt request. This operation is shown in Figure 6.12.

**Receive Operation**

The receive operation is the counterpart of the transmit operation. Transmission is initiated by the UART, but reception is initiated by some external device, such as another UART.

The UART senses the start bit of a new byte when the din input goes low. When this happens, the model waits for another half clock to see if din is still low, and if so, the model starts receiving the byte. The model sets status to RECEIVING and starts sampling din at each clock.

```
always @do_transmit begin : transmit_block
    integer i;
    reg[7:0] data;
    reg[7:0] status;

    // Set the TRANSMITTING bit in the status register
    ...
    oclkreg = 0;
    for (i = 0; i < 8; i = i + 1) begin
       @(posedge xmt_clock)
       dout = data[i];
    end

    // Reset the TRANSMITTING bit in the status register
    // and set the DONE_XMT bit
    ...
    int = 1;
end
```

**Figure 6.12** Transmit operation

```
always @(negedge din) begin : receive_block
    integer i;
    reg[7:0] data;
    reg[7:0] status;

    // wait for half clock
    iclkreg = 0;
    while (iclkreg != divisor / 2) @clkin ;
    if (din != 0) disable receive_block;

    // Start receiving
    // Set the RECEIVING bit in the status register
    ...
    iclkreg = 0;
    for (i = 0; i < 8; i = i + 1) begin
       @(posedge rcv_clock)
       data[i] = din;
    end
    // Reset the RECEIVING bit in the status register
    // and set the DONE_RCV bit
    ...
    int = 1;
end
```

**Figure 6.13** Receive block

When done the model resets the status to DONE_RCV and raises the interrupt flag. This is shown in Figure 6.13.

## Testing the Dual UART Chip

After writing the structural model of the dual UART and the functional model of the single UART, we need to develop a driver module to test the chip.

Figure 6.14 shows the top module for the chip-level simulation. It has the following parts: first, the dual UART module itself is instantiated; second, the external clock is modeled; third, some initialization takes place; fourth, some of the testing operations are coded as small tasks; fifth, the test itself is coded as an "initial" block which strings together some of the tasks into a single test; finally, a loop observes the interrupt output from the UART and displays a message.

The following piece of code is composed of several small tasks which can be strung together into a more complex sequence or which can be issued interactively. The first task, do_reset, issues a reset sequence by toggling the reset input. The next two tasks, writereg and readreg, initiate write and read sequences to an internal register r. They manipulate the a, cs_, wr_, and dbus inputs as required by the waveforms of Figure 6.15. The fourth task, receivebyte, initiates transmission of a byte to the UART.

The actual test consists of a reset sequence, the setting of the baud rate, a transmission of a byte, and a reception of a byte, as shown in Figure 6.16.

## Implementation of the Single UART

The functional model described in the previous sections is not concerned with the efficiency, or even the feasibility, of implementation. Now, we develop a model that is more closely related to the gate-level implementation and is more amenable to synthesis.

There are two main differences between the functional model and the implementation-oriented model of the UART. One difference is that, in the latter, the register file is not implemented as an array; instead, each

```
module topu;
  reg [7:0] dbus_reg;
  wire [7:0] dbus = dbus_reg;
  reg[2:0] a;
  reg reset, rd_, wr_;
  reg cs_0, cs_1, din0, din1, clkin;

  wire int0, int1, dout0, dout1, clko0, clko1;

  // Instantiate the dual UART module
  dual_uart dual_uart1 (
     dbus, a, reset, rd_, wr_,
     cs_0, cs_1, din0, din1, clkin,

     int0, int1, dout0, dout1, clko0, clko1
  );

  // Generate the external clock
  parameter halfcycle = 2;
  initial clkin = 0;
  always #halfcycle clkin = ~clkin;

  initial begin
     // Set all the internal registers to 0
     ...
     #250 $finish;
  end

  // All the tasks and functions of the driver
  // are described in the following sections

  // task do_reset . . .
  // task receivebyte . . .
  // task readreg . . .
  // test_block . . .

endmodule
```

**Figure 6.14** UART test module

register is separate and is accessed by decoding the address lines a[2:0]. This enables us to set and reset single bits in the status register.

Another major difference between the two models is the replacement of the counters by shift registers. When data is transmitted or received, we need to count eight clocks, one clock per bit. In the functional model, two counters are used for this purpose. Here, we concatenate a tag to the end of the data to shift, and we detect the end of transmission when the tag has been shifted to the end of the shift register.

```
task resettask;
  begin
     reset = 1;
     #1 reset = 0;
  end
endtask

task writereg;
input [2:0] r;
input [7:0] val;
input cs;
begin
   a = r;
   if (cs) cs_1 = 0;
   else cs_0 = 0;
   dbus_reg = val;
   #1 wr_ = 0;
   #1 wr_ = 1;
   #1 dbus_reg = 8'hzz;
   if (cs) cs_1 = 1;
   else cs_0 = 1;
end
endtask

task receivebyte;
input [7:0] val;
integer i;
begin
   @(posedge clko0)
   #halfcycle
   din0 = 0;
   for (i = 0; i < 8; i = i + 1) begin
      @(posedge clko0)
      din0 = val[i];
   end
end
endtask

task readreg;
input [2:0] r;
input cs;
begin
   a = r;
   if (cs) cs_1 = 0;
   else cs_0 = 0;
   #1 rd_ = 0;
   #1 $display ("readreg, val = %h", dbus);
   #1 rd_ = 1;
   if (cs) cs_1 = 1;
   else cs_0 = 1;
end
endtask
```

**Figure 6.15** Tasks of the UART test module

```
initial begin : test_block
    integer i;

    #1
    resettask;

    // Set the clock divider to 3
    writereg (2, 3, 0);
    writereg (3, 0, 0);

    // Start transmitting (byte 0F)
    writereg (0, 8'h0f, 0);
    wait (int0);
    #1

    // Clear the interrupt and transmit again (byte AA)
    readreg (7, 0);
    #1
    writereg (0, 8'haa, 0);
    wait (int0);
    #1

    // Clear the interrupt and receive
    readreg (7, 0);

    // Receive a byte
    receivebyte (8'hc7);
    wait (int0);
    #1
    readreg (4, 0);
    #1

    // Clear the interrupt
    readreg (7, 0);
end
always @(posedge int0) begin
    $display
     ("Received interrupt from int0,clearing at time %0d",
     $time);
end
always @(int0 or dout0 or clko0) begin
    $display
     ("int0,dout0,clko0=%b %b %b",int0,dout0,clko0);
end
```

**Figure 6.16** Test block to test single_uart module

## Modeling Asynchronous I/O: UART

```
module uart (
    reset, dbus, a, rd_, wr_,
    cs_, din, clkin,
    int, dout, clko
);
// Input/output declarations
...

// Parameter declarations
...

reg[7:0] divmsb_reg, divlsb_reg;
reg[7:0] xmit_reg, recv_reg;
wire[15:0] divisor = { divmsb_reg, divlsb_reg };
wire[15:0] halfdivisor = { 1'b0, divisor[15:1] };
reg[15:0] oclkreg, iclkreg;
reg xmt_clock, rcv_clock;
event do_transmit;

reg transmitting;
reg receiving;
reg xmt_done;
reg rcv_done;
reg rcv_tag, xmt_tag;

wire[7:0] status =
    {4'hx,transmitting,receiving,xmt_done,rcv_done };

// Reset operation.
always @(negedge reset) begin : reset_block
    // Set all the internal registers to 0
    ...
end

// Clock generator block.
  // ...

// Read operation
function [7:0] outdbus;
input[2:0] a;
begin
    case (a)
    XMITDT_ADDR: outdbus = xmit_reg;
    STATUS_ADDR: outdbus = status;
    DIVLSB_ADDR: outdbus = divlsb_reg;
    DIVMSB_ADDR: outdbus = divmsb_reg;
    RECVDT_ADDR: outdbus = recv_reg;
    default: ;
    endcase
end
endfunction                              continued
```

**Figure 6.17** Implementable model for the single UART

```verilog
   assign int = xmt_done | rcv_done;

   always @(posedge reading) begin
      dbus_reg = outdbus (a);
      @(posedge rd_)
      dbus_reg = 8'hzz;
      if (a == CLRINT_ADDR) begin : read_block
         xmt_done = 0;
         rcv_done = 0;
      end
   end

   // Write operation.
   always @(posedge writing) begin
      @(posedge wr_)
      case (a)
      XMITDT_ADDR: begin
         xmit_reg = dbus;
         transmitting = 1;
      end
      STATUS_ADDR: { transmitting, receiving, xmt_done, rcv_done } = dbus[3:0];
      DIVLSB_ADDR: divlsb_reg = dbus;
      DIVMSB_ADDR: divmsb_reg = dbus;
      RECVDT_ADDR: recv_reg = dbus;
      default: ;
      endcase
   end

   // Transmit operation.
   always @(posedge transmitting) begin : transmit_block
      xmt_tag = 1;
      oclkreg = 0;
   end

   always @(posedge xmt_clock) if (transmitting) begin
      // right shift the xmt_reg to dout
      { xmt_tag, xmit_reg, dout } =
         { 1'b0, xmt_tag, xmit_reg };
   end

   wire done_transmit = { xmt_tag, xmit_reg } == 9'b000000001;

   always @(posedge done_transmit) begin
      transmitting = 0;
      xmt_done = 1;
   end

   // Receive operation.
   always @(negedge din) begin : receive_block
```
**continued**

**Figure 6.17** Implementable model for the single UART (continued)

## Modeling Asynchronous I/O: UART

```
    // wait for half clock
    ...

    // Start receiving
    receiving = 1;
    iclkreg = 0;
    { recv_reg, rcv_tag } = 9'b100000000;
end

wire done_receive = (rcv_tag == 1);

always @(posedge rcv_clock)
    if (receiving)
        // right shift dout into the receive register
        { recv_reg, rcv_tag } = { din, recv_reg };

always @(posedge done_receive) begin
    receiving = 0;
    rcv_done = 1;
end
endmodule
```

**Figure 6.17** Implementable model for the single UART (continued)

A skeleton of the implementation-oriented model is shown in Figure 6.17.

## Summary

In this chapter we demonstrated some techniques of modeling asynchronous I/O using a UART as an example. Among the points to note are the use of a combination of structural and behavioral models, the handling of asynchronous bus transactions, the modeling of bidirectional ports, the use of tasks and functions to improve readability, the use of continuous assignments to model combinational functions, and the construction of complex tasks from simple tasks.

The complete source code for the dual and single UART model is given on the following pages.

**Model for the Single and Dual UART**

```verilog
module top;
   reg [7:0] dbus_reg;
   wire [7:0] dbus = dbus_reg;
   reg [2:0] a;
   reg reset, rd_, wr_;
   reg cs_0, cs_1, din0, din1, clkin;

   wire int0, int1, dout0, dout1, clkout0, clkout1;

   // Instantiate the dual uart module
   dual_uart dual_uart1 (
      dbus, a, reset, rd_, wr_,
      130cs_0, cs_1, din0, din1, clkin,
      int0, int1, dout0, dout1, clkout0, clkout1
   );

   // Generate the external clock
   parameter halfcycle = 2;
   initial clkin = 0;
   always #halfcycle clkin = ~clkin;

   initial begin
      reset = 0;
      rd_ = 1;
      wr_ = 1;
      cs_0 = 1;
      cs_1 = 1;
      din0 = 1;
      din1 = 1;
      a = 0;
      #250 $finish;
   end

   task resettask;
   begin
      reset = 1;
      #1 reset = 0;
   end
   endtask

   task writereg;
   input [2:0] r;
   input [7:0] val;
   input cs;
   begin
      $display ("writereg, r = %0d, val = %0d, cs = %b",
             r, val, cs);
      a = r;
      if (cs) cs_1 = 0;
      else cs_0 = 0;
      dbus_reg = val;
      #1 wr_ = 0;
      #1 wr_ = 1;
      #1 dbus_reg = 8'hzz;
```

## Modeling Asynchronous I/O: UART

```verilog
      if (cs) cs_1 = 1;
      else cs_0 = 1;
   end
endtask

task receivebyte;
input [7:0] val;
integer i;
begin
   $display ("receivebyte, val = %0d", val);
   @(posedge clkout0)
   #halfcycle
   din0 = 0;
   for (i = 0; i < 8; i = i + 1) begin
      @(posedge clkout0)
      din0 = val[i];
      // $display ("bit = %b at time %0d", din0, $time);
   end
end
endtask

task readreg;
input [2:0] r;
input cs;
begin
   $display ("readreg, r = %0d, cs = %b", r, cs);
   a = r;
   if (cs) cs_1 = 0;
   else cs_0 = 0;
   #1 rd_ = 0;
   #1 $display ("readreg, val = %h", dbus);
   #1 rd_ = 1;
   if (cs) cs_1 = 1;
   else cs_0 = 1;
end
endtask

initial begin : test_block
   integer i;

   #1
   resettask;
   // Set the clock divider to 3
   writereg (2, 3, 0);
   writereg (3, 0, 0);
   // Start transmitting (byte 0F)
   writereg (0, 8'h0f, 0);
   wait (int0);
   #1
   // Clear the interrupt and transmit again (byte AA)
   readreg (7, 0);
   #1
   writereg (0, 8'haa, 0);
   wait (int0);
   #1
   // Clear the interrupt and receive
```

```verilog
        readreg (7, 0);
        // Receive a byte
        receivebyte (8'hc7);
        wait (int0);
        #1
        readreg (4, 0);
        #1
        // Clear the interrupt
        readreg (7, 0);
    end

    always @(posedge int0) begin
      $display("Received interrupt from int0, clearing at %0d",
         $time);
    end

    always @(int0 or dout0 or clkout0) begin
       $display ("int0, dout0, clkout0 = %b %b %b",
          int0, dout0, clkout0);
    end

    endmodule

/* =========================================*/
    module dual_uart (
       dbus, a, reset, rd_, wr_,
       cs_0, cs_1, din0, din1, clkin,
       int0, int1, dout0, dout1, clkout0, clkout1
    );
    inout[7:0] dbus;
    wire[7:0] dbus;
    input[2:0] a;
    wire[2:0] a;
    input reset, rd_, wr_;
    wire reset, rd_, wr_;
    input cs_0, cs_1, din0, din1, clkin;
    wire cs_0, cs_1, din0, din1, clkin;
    output int0, int1, dout0, dout1, clkout0, clkout1;
    wire int0, int1, dout0, dout1, clkout0, clkout1;

    uart u0 (
       reset, dbus, a, rd_, wr_,
       cs_0, din0, clkin,
       int0, dout0, clkout0
    );
    uart u1 (
       reset, dbus, a, rd_, wr_,
       cs_1, din1, clkin,
       int1, dout1, clkout1
    );
    endmodule

    // The functional model for the single UART
    module uart (
       reset, dbus, a, rd_, wr_,
       cs_, din, clkin,
```

## Modeling Asynchronous I/O: UART

```verilog
        int, dout, clkout
);
inout[7:0] dbus;
reg [7:0] dbus_reg;
wire[7:0] dbus = dbus_reg;
input[2:0] a;
wire[2:0] a;
input reset, rd_, wr_, cs_, din, clkin;
wire reset, rd_, wr_, cs_, din, clkin;
output int, dout, clkout;
reg int, dout, clkout;

// The addresses of the internal registers
parameter
    XMITDT_ADDR = 0,
    STATUS_ADDR = 1,
    DIVLSB_ADDR = 2,
    DIVMSB_ADDR = 3,
    RECVDT_ADDR = 4,
    CLRINT_ADDR = 7;

// The internal state
parameter
    TRANSMITTING = 0,
    RECEIVING = 1,
    DONE_XMT = 2,
    DONE_RCV = 3;

reg[7:0] regfile [0:7];
wire[15:0]divisor =
    {regfile[DIVMSB_ADDR],regfile[DIVLSB_ADDR] };
reg[15:0] oclkreg, iclkreg;
reg xmt_clock, rcv_clock;
event do_transmit;

integer lines;

initial begin
    lines = 0;
    oclkreg = 0;
    iclkreg = 0;
    $monitor(
    "%m:r[0]=%h,r[1]=%h,r[2]=%h,r[3]=%h,r[4]=%h,time=%0d"
        ,regfile[0],regfile[1],regfile[2]
        ,regfile[3],regfile[4],$time);
end

/*
always @(dbus or a or reset or rd_ or wr_ or
    cs_ or din or clkin) begin
    if ((lines % 15) == 0)
        $display(
        "module dbus a reset rd_ wr_ cs_ din clkin time");
    lines = lines + 1;
    $display(
    "%m: %h    %0d    %b    %b    %b    %b    %b    %b    %0d"
```

183

```
              ,dbus, a, reset, rd_, wr_, cs_, din, clkin, $time);
        end
   */
   /*
   Reset operation. On the negative edge of reset, set all the
   registers to 0 and tristate the outputs.
   */
        always @(negedge reset) begin : reset_block
           integer i;

           disable receive_block;
           disable transmit_block;
           oclkreg = 0;
           iclkreg = 0;
           int = 0;
           dbus_reg = 8'hzz;
           for (i = 0; i < 8; i = i + 1)
              regfile[i] = 0;
        end

/*
Clock generator block. The external clock is divided by the
divisor (from regfile[3] and regfile[2]), and a divided clock
is generated for receive and transmit. The receive and
transmit clocks are of the same frequency but may be
shifted with respect to one another. The two clocks are
initialized by loading iclkreg and oclkreg.
*/
        always @(posedge clkin) begin
           oclkreg = oclkreg + 1;
           iclkreg = iclkreg + 1;
           fork
           if (oclkreg >= divisor) begin
              oclkreg = 0;
              xmt_clock = 1;
              clkout = 1;
              #1
              clkout = 0;
              xmt_clock = 0;
           end
           if (iclkreg >= divisor) begin
              iclkreg = 0;
              rcv_clock = 1;
              #1
              rcv_clock = 0;
           end
           join
        end

/*
Read operation. On the negative edge of rd_, put the
appropriate register on the bus, and on the positive edge of
rd_, tristate the bus. In addition, if the address being read
is the status, reset the DONE_XMT and DONE_RCV bits of the
status.
*/
```

## Modeling Asynchronous I/O: UART

```verilog
      always @(negedge rd_) if (~cs_) begin
         dbus_reg = regfile[a];
         @(posedge rd_)
         dbus_reg = 8'hzz;
         if (a == CLRINT_ADDR) begin : read_block
            reg[7:0] status;
            status = regfile[STATUS_ADDR];
            int = 0;
            status[DONE_XMT] = 0;
            status[DONE_RCV] = 0;
            regfile[STATUS_ADDR] = status;
         end
      end

/*
Write operation. Load the appropriate register with the value
from the data bus. In addition, if the address indicates a
new data byte to transmit, start transmitting.
*/
      always @(negedge wr_) if (~cs_) begin
         @(posedge wr_)
         regfile[a] = dbus;
         if (a == XMITDT_ADDR) ->do_transmit;
      end

/*
Transmit operation. Set the status to TRANSMITTING, reset the
transmit clock, and start transmitting. When done set the
status to DONE_XMT and raise the interrupt.
*/
      always @do_transmit begin : transmit_block
         integer i;
         reg[7:0] data;
         reg[7:0] status;

//                   $display ("In do_transmit");
         status = regfile[STATUS_ADDR];
         status[TRANSMITTING] = 1;
         regfile[STATUS_ADDR] = status;
         data = regfile[XMITDT_ADDR];
         oclkreg = 0;
         for (i = 0; i < 8; i = i + 1) begin
            @(posedge xmt_clock)
//                   $display ("In xmt_clock");
            dout = data[i];
         end
         status = regfile[STATUS_ADDR];
         status[TRANSMITTING] = 0;
         status[DONE_XMT] = 1;
         regfile[STATUS_ADDR] = status;
         int = 1;
      end

/*
Receive operation. On the negative edge of din, wait for
another half clock to see if din is still low, and if yes,
```

```
start receiving. Set the status to RECEIVING, and start
sampling din at each clock. When done reset the status to
DONE_RCV and raise the interrupt flag.
*/
   always @(negedge din) begin : receive_block
      integer i;
      reg[7:0] data;
      reg[7:0] status;
      /* wait_half_clock */
      iclkreg = 0;
      while (iclkreg != divisor / 2) @clkin ;
      if (din != 0) disable receive_block;
      /* Start receiving */
      status = regfile[STATUS_ADDR];
      status[RECEIVING] = 1;
      regfile[STATUS_ADDR] = status;
      iclkreg = 0;
      for (i = 0; i < 8; i = i + 1) begin
         @(posedge rcv_clock)
         data[i] = din;
      end
      regfile[RECVDT_ADDR] = data;
      status = regfile[STATUS_ADDR];
      status[RECEIVING] = 0;
      status[DONE_RCV] = 1;
      regfile[STATUS_ADDR] = status;
      int = 1;
   end

endmodule


//      The implementation model for the single UART
module uart (
   reset, dbus, a, rd_, wr_,
   cs_, din, clkin,
   int, dout, clkout
);
inout[7:0] dbus;
reg[7:0] dbus_reg;
wire[7:0] dbus = dbus_reg;
input[2:0] a;
wire[2:0] a;
input reset, rd_, wr_, cs_, din, clkin;
wire reset, rd_, wr_, cs_, din, clkin;
wire reading = ~cs_ & ~rd_;
wire writing = ~cs_ & ~wr_;

output int, dout, clkout;
wire int;
reg clkout, dout;

parameter
   XMITDT_ADDR = 0,
   STATUS_ADDR = 1,
   DIVLSB_ADDR = 2,
```

## Modeling Asynchronous I/O: UART

```verilog
         DIVMSB_ADDR = 3,
         RECVDT_ADDR = 4,
         CLRINT_ADDR = 7;

      parameter
         TRANSMITTING = 0,
         RECEIVING = 1,
         DONE_XMT = 2,
         DONE_RCV = 3;

      reg[7:0] divmsb_reg, divlsb_reg;
      reg[7:0] xmit_reg, recv_reg;
      wire[15:0] divisor = { divmsb_reg, divlsb_reg };
      wire[15:0] halfdivisor = { 1'b0, divisor[15:1] };
      reg[15:0] oclkreg, iclkreg;
      reg xmt_clock, rcv_clock;
      event do_transmit;

      reg transmitting;
      reg receiving;
      reg xmt_done;
      reg rcv_done;
      reg rcv_tag, xmt_tag;

      wire[7:0] status =
         { 4'hx, transmitting, receiving, xmt_done, rcv_done };

/*
Reset operation. On the negative edge of reset, set all the
registers to 0 and tristate the outputs.
*/
      always @(negedge reset) begin : reset_block
         integer i;
         oclkreg = 0;
         iclkreg = 0;
         dbus_reg = 8'hzz;
         xmit_reg = 0;
         transmitting = 0;
         receiving = 0;
         xmt_done = 0;
         rcv_done = 0;
         divlsb_reg = 0;
         divmsb_reg = 0;
         recv_reg = 0;
      end

/*
Clock generator block. The external clock is divided by the
divisor (from regfile[3] and regfile[2]), and a divided clock
is generated for receive and transmit. The receive and
transmit clocks are of the same frequency but may be shifted
with respect to one another. The two clocks are initialized by
loading iclkreg and oclkreg.
*/
      always @(posedge clkin) begin
         oclkreg = oclkreg + 1;
```

```verilog
      iclkreg = iclkreg + 1;
      fork
         if (oclkreg >= divisor) begin
            oclkreg = 0;
            xmt_clock = 1;
            clkout = 1;
            #1
            clkout = 0;
            xmt_clock = 0;
         end
         if (iclkreg >= divisor) begin
            iclkreg = 0;
            rcv_clock = 1;
            #1
            rcv_clock = 0;
         end
      join
   end
```

/*
Read operation. On the negative edge of rd_, put the
appropriate register on the bus, and on the positive edge of
rd_, tristate the bus. In addition, if the address being read
is the status, reset the DONE_XMT and DONE_RCV bits of the
status.
*/

```verilog
   function [7:0] outdbus;
   input[2:0] a;
   begin
      case (a)
      XMITDT_ADDR: outdbus = xmit_reg;
      STATUS_ADDR: outdbus = status;
      DIVLSB_ADDR: outdbus = divlsb_reg;
      DIVMSB_ADDR: outdbus = divmsb_reg;
      RECVDT_ADDR: outdbus = recv_reg;
      default: ;
      endcase
   end
   endfunction

   assign int = xmt_done | rcv_done;

   always @(posedge reading) begin
      dbus_reg = outdbus (a);
      @(posedge rd_)
      dbus_reg = 8'hzz;
      if (a == CLRINT_ADDR) begin : read_block
         xmt_done = 0;
         rcv_done = 0;
      end
   end
```

/*
Write operation. Load the appropriate register with the value
from the data bus. In addition, if the address indicates a

## Modeling Asynchronous I/O: UART

```
new data byte to transmit, start transmitting.
*/
   always @(posedge writing) begin
      @(posedge wr_)
      case (a)
      XMITDT_ADDR: begin
         xmit_reg = dbus;
         transmitting = 1;
      end
      STATUS_ADDR:
       {transmitting,receiving,xmt_done,rcv_done}=dbus[3:0];
      DIVLSB_ADDR: divlsb_reg = dbus;
      DIVMSB_ADDR: divmsb_reg = dbus;
      RECVDT_ADDR: recv_reg = dbus;
      default: ;
      endcase
   end

/*
Transmit operation. Set the status to TRANSMITTING, reset the
transmit clock and start transmitting. When done set the
status to XMT_DONE and raise the interrupt.
*/
   always @(posedge transmitting) begin : transmit_block
      xmt_tag = 1;
      oclkreg = 0;
   end

   always @(posedge xmt_clock) if (transmitting) begin
      // right shift the xmt_reg to dout
      { xmt_tag, xmit_reg, dout } = { 1'b0, xmt_tag, xmit_reg };
   end

   wire done_transmit = { xmt_tag, xmit_reg } == 9'b000000001;

   always @(posedge done_transmit) begin
      transmitting = 0;
      xmt_done = 1;
   end

/*
Receive operation. On the negative edge of din, wait for
another half clock to see if din is still low, and if yes,
start receiving. Set the status to RECEIVING, and start
sampling din at each clock. When done reset the status to
DONE_RCV and raise the interrupt fllag.
*/
   always @(negedge din) begin : receive_block
      /* wait_half_clock */
      iclkreg = 0;
      while (iclkreg != halfdivisor) @clkin ;
      if (din != 0) disable receive_block;
      /* Start receiving */
      receiving = 1;
      iclkreg = 0;
      { recv_reg, rcv_tag } = 9'b100000000;
```

189

```
   end

   wire done_receive = (rcv_tag == 1);

   always @(posedge rcv_clock)
      if (receiving)
         // right shift dout into the receive register
         { recv_reg, rcv_tag } = { din, recv_reg };

   always @(posedge done_receive) begin
      receiving = 0;
      rcv_done = 1;
   end

   endmodule//
```

CHAPTER

# 7

# Verilog HDL for Synthesis

## Introduction

In the previous chapters we introduced Verilog HDL and showed how it can be used in different ways to support top-down hierarchical design. In this chapter we cover the basics of synthesis, discuss how Verilog may be used for synthesis, and describe how modeling for synthesis affects the coding style, the design organization, and partitioning.

**What is Synthesis?**

Synthesis is the process of taking a high level design description and implementing it using library components. In our case, the source language is a Verilog subset, and the target library is a vendor specific set of components. Typically, the library is provided by an ASIC vendor; but it can also consist of programmable device building blocks. However, merely converting Verilog descriptions into gates is not sufficient. Of critical importance is the process of design optimization.

Synthesis tools provide facilities for expressing design constraints such as area, speed, and power. Depending upon the design requirements, the optimization process often involves multiple synthesis runs for each module, using a combination of these constraints to produce a design as close to optimal as possible.

Different terminologies have been used to define various levels of synthesis. Figure 7.1 illustrates different levels of synthesis, various operations associated with each level, and different representations at each level. Precise definitions do not exist for some of the highest levels of synthesis such as system-level synthesis. This is so because at these abstract levels the synthesis process is not well defined.

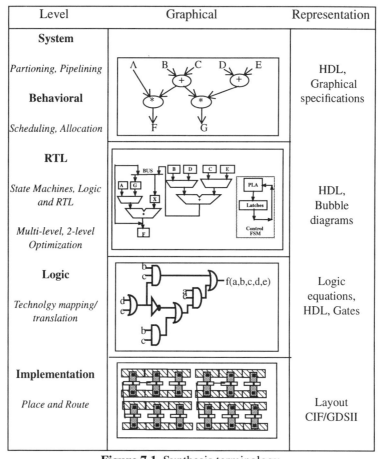

**Figure 7.1** Synthesis terminology

Behavioral, or high level, synthesis refers to an abstract level where we describe an algorithm, a synthesis tool designs (or selects) an architecture, and then the tool allocates and schedules operations using this architecture to implement the algorithm.

Register Transfer Level (RTL) is a description at a level that provides complex operators that operate on multi-bit register variables, signals and constants. RTL synthesis assumes that the architecture for the design has been defined and the allocation and scheduling of operations to implement an algorithm have been determined. The description is in the form of a data-flow graph, which may be represented in a hardware description language and requires that we describe the design in terms of registers and operations.

Finite state machine (FSM) synthesis is often considered a part of RTL synthesis and involves the translation of state-transition and encoding information into registers and combinational logic functions.

Logic synthesis deals with combinational circuits and requires that we provide a set of Boolean, or logic, equations describing the circuit. The description requires detailed logical representation of the functions and is considered the lowest level of synthesis. However, it is worth noting that logic synthesis operates at a higher level than transistor and gate level netlists.

Currently, commercial synthesis tools cannot synthesize high level behavioral descriptions, and the limitations of these tools often require restructuring of the Verilog modules. Commercial synthesis tools operate at the RTL and Boolean logic levels. Since we are interested in studying how Verilog can be used for synthesis, we assume that the architecture for the design has already been defined. In the rest of this chapter we will use the term HDL synthesis to mean RTL and logic synthesis.

**HDL Synthesis**

HDL synthesis converts an RTL description into a gate level design. RTL synthesis tools assume a synchronous design methodology where the design is at all times in one of a finite number of states. All changes in states are triggered on the active edge of a clock. An exception to the design methodology may provide an asynchronous reset signal to put the design in a known initial state. The state is determined by the value stored in a finite number of clocked storage elements, which correspond to flip-flops in the synthesized logic.

Synthesis tools assume that all inputs to the design aside from the clock(s) and asynchronous reset are derived from storage elements triggered by the clock or are primary inputs to the design. As a result, from a functional point of view, as long as all combinational logic settles within the clock period, it can be assumed that the combinational logic evaluates in zero time. This simplifies the task of writing Verilog models by eliminating the need to consider technology-dependent timing information.

Figure 7.2 illustrates the synthesis process. It has three distinct phases:

1. **Design Compilation:** In this first phase, the synthesis tool parses and compiles the RTL description into generic technology-independent logic equations.

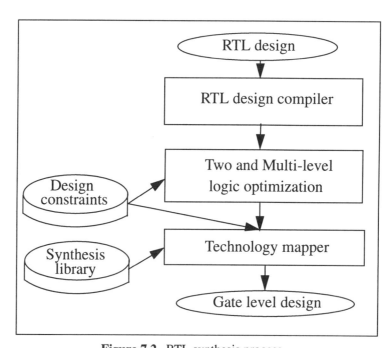

**Figure 7.2** RTL synthesis process

2. *Logic Optimization:* In the second phase, the logic goes through technology independent optimization which aims at minimizing the number of terms in the logic equations.

   At this stage, some of the design constraints are evaluated and the optimization tool chooses a design that best meets these constraints. The optimization occurs in two phases:

   *2.1. Multi-level logic optimization*

   In multi-level optimizations logic equations are simplified, resulting in improved performance and sharing of complex multi-bit operators.

   *2.2 2-level logic optimization*

   In 2-level optimization, Boolean equations are optimized to two-level AND-OR representations.

3. *Technology Mapping:* In the third phase, the synthesized logic is mapped to a gate level design using an ASIC vendor library. In this stage a technology mapper reads the process technology library and uses the design constraints, such as area, to convert the logic equations into a gate level netlist.

   Technology translation is an extension to the technology mapping process and provides a capability of reading gate level netlists from one technology and converting them to another.

**Synthesis Benefits**

The use of Verilog HDL synthesis tools can eliminate or reduce effort at several stages. Here, we will discuss some important benefits of synthesis.

1. Synthesis reduces the time to generate gates from an HDL description. A commitment to top-down design generally implies that a design will be described and simulated at a high level before being mapped into gates. In general, the synthesis process will generate a gate level design faster than manual implementation.

2. Synthesis provides a fast mechanism for re-targeting designs from one technology to another. As fabrication processes improve, more gates can be accommodated and the HDL descriptions of multiple ASICs can easily be combined to fully utilize increased die sizes.

3. Synthesis reduces the gate level design debugging effort. Given accurate target ASIC libraries, synthesis guarantees to generate a design functionally equivalent to the source module. Manual translation of an RTL module into gates is error prone. Typical errors are phase inversions and erroneous connections in the schematic capture process. Since these errors do not occur with logic synthesis, they do not need to be painstakingly found during simulation.

4. Synthesis provides consistency between the HDL and gate versions of a design. Without synthesis, keeping such consistency represents a major design management challenge because any change in the design needs to be propagated manually to both versions.

5. RTL simulations permit a higher level of design verification, where errors are detected prior to the actual gate level implementation, thereby shortening the design cycle.

6. In general, synthesis may improve the design by producing a more efficient gate level implementation than a designer could create by hand. The design could be better in terms of speed, gate utilization, routability, power consumption, or some other metric. There is no guarantee that synthesis will in fact produce a better design. Synthesis may still be preferred even if an expert designer could do a better job because of its design time, technology re-targeting, design integrity, and complexity management advantages.

Most fundamentally, synthesis will make top-down design really work by allowing low-level designs to be generated programmatically from a single high-level representation. Thus, in the top-down design method, the HDL description forms the main design description and documentation.

# Verilog HDL for Synthesis

**Practical Considerations**

The benefits outlined in the previous section demonstrate why synthesis can be a desirable addition to a design environment. Once synthesis is available to designers, some interesting practical questions arise. In this section we briefly discuss some of these issues.

The most obvious question is whether we can trust the synthesized logic. The design of HDL synthesis tools are intended to assure that the resulting gate level design will be functionally equivalent to the input description. Still, we recommend that the designer verify the synthesis output.

Another practical issue is the combination of synthesized and non synthesized modules in a single chip design. We may choose to reuse a module from an old design, or perhaps to design a specialized function at the gate level. Such designs can be performed in Verilog, either by using its gate level functions or instantiating cells from the target ASIC library.

A gate level design captured using schematic capture tools can be merged with the synthesized design. The gate level modules may be imported into the chip during synthesis and linked into the design. It is also possible to merge synthesized and manually designed modules using symbols for the synthesized logic and gate level schematics. Most synthesis tools provide the capability of creating symbols or provide schematic output capability.

## Synthesis Design Method

Synthesis can use the top-down Verilog HDL design methodology we described in the previous chapters. Figure 7.3 illustrates a design flow that includes five different phases of operation:

1. Design at Register Transfer Level
2. Functional verification of the RTL design
3. Gate level implementation
4. Logic and timing verification
5. Physical implementation

**Figure 7.3** Synthesis design flow

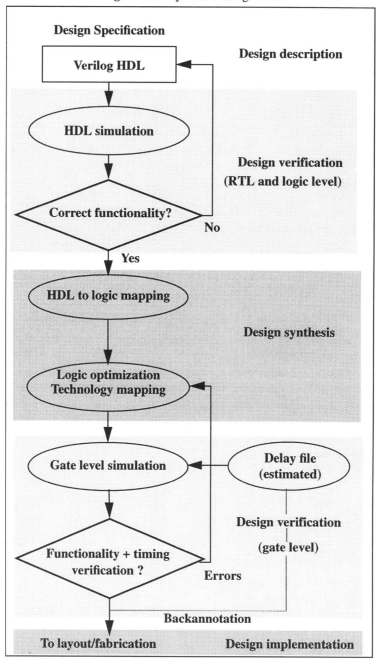

## Design at the Register Transfer Level

The first phase of the design methodology is the hierarchical decomposition of the design from an abstract specification to a high level design description. To achieve this, we work through the design specification hierarchy and then map this hierarchy into the Verilog RTL or gate level modules. This process can be viewed as the creation of a hierarchy of block diagrams, where each lower level reveals more details of the design than the level above. For example, we may start by defining a computer as interconnections between a memory block, a processor block, and an input/output block. The memory block may then be decomposed into blocks of memory banks, data paths, and control logic. Each memory bank may then be decomposed into its individual memory chips.

Verilog facilitates this style of design by providing hierarchical modules in much the same way that a programming language provides functions and procedures. Each function or block diagram can be described as one or more Verilog modules. The language constructs encourage a top-down hierarchical decomposition. Interconnection of blocks at the same level of hierarchy is done by instantiating modules at the same level. This process of design decomposition into hierarchical Verilog modules yields a description of the design at the RTL.

## Functional Verification

The second phase of the design method is the verification of this RTL design. Functional vectors are used to stimulate the design. These vectors generally verify the inter-module communications and associated protocols of the device that need to be described. The vectors can be generated using procedural and tabular vectors, or, more commonly, by writing behavioral models in Verilog.

The recommended practice is to instantiate into a test module the Verilog module describing the design. The test module contains behavioral code that applies stimulus to the design-under-test. The purpose of the test module is to confirm whether the design is functionally correct. If it is, the design can be "removed" from the test module and migrated to the next step in the design flow. Only the design itself is passed to the synthesis tool — not the module that was used to test it. Test modules sit outside the design and are used solely for verifying the correctness of the design — before the design is synthesized. For example, Figure 7.4 illustrates the case where we may use behavioral models for the I/O block, memory, and processor. This approach also permits the reuse of the same stimulus and behavioral models for gate level verification in the final phase.

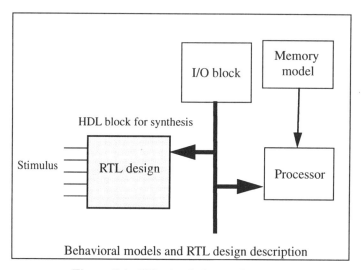

**Figure 7.4**  RTL simulation environment

### Gate Level Implementation

The third phase of this design flow synthesizes the RTL design into gate level primitives of a specific ASIC library. Synthesis tools are used primarily at this level of the design. Area, power, and speed constraints are used to obtain an optimal design. The implementation of the design at the gate level is then used to build the actual chip, board, or system.

### Logic and Timing Verification

The fourth phase is the functional and timing verification of the synthesized logic. Figure 7.5 illustrates a typical design verification strategy. HDL synthesis tools generate a gate level design that is functionally equivalent to the RTL description of each Verilog module in the design.

The timing performance of the synthesized logic must be examined to verify that the circuit meets its timing specifications as well as its functionality. In general, layout-induced loading effects (caused by metal interconnect capacitance) may cause a circuit to fail to operate correctly at the specified clock frequencies. At this step in the design flow the physical layout of the circuit has not been created, so estimations of interconnect delays that determine the propagation delays must be injected into the models. The behavioral test models and the stimulus used for functional verification of the RTL description can be used during this verification process. If errors are discovered during the simulation, we may wish to modify the RTL description or change the optimization constraints until the design specifications are satisfied.

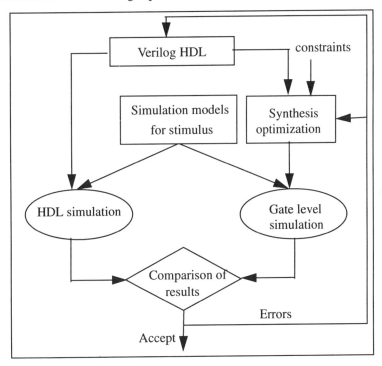

**Figure 7.5** Design verification

**Physical Implementation**

The resulting gate level design is a Verilog netlist that you can use to build the physical implementation and create the mask-level description of the design. The gate level designs are transferred to the polygon-level using silicon compilers and physical synthesis tools, such as module generators, or by routers.

Synthesis tools do not place and route the components and, as a result, they can only estimate the propagation delays caused by interconnect wires. After layout, more accurate timing information is available which we can use to do post-layout timing analysis and detect any timing violations that were introduced due to long metal paths. An iterative process between the synthesis tool and the physical design tools is used to correct any timing violations. This is done by readjusting the design constraints for each of the successive synthesis runs.

## Design Style for Verilog Synthesis

Currently available synthesis tools cannot efficiently synthesize the full Verilog language. When writing Verilog description for synthesis, we need to consider the constraints imposed by the synthesis tools. This implies using only a subset of the language and also using some coding guidelines. Personal preferences also play an important role in coding style.

The basic constraint in writing Verilog for synthesis is using a language subset acceptable by the synthesis tool for the modules to be synthesized. This subset may vary from vendor to vendor. All synthesis examples in this book conform to the requirements of most current synthesis tools. For specific constructs and their implications, refer to the particular synthesis tool's reference manuals.

Next we provide a list of generic rules which can be used to ensure that a Verilog module is synthesizable. For a list of Verilog constructs, characterized as fully synthesizable, partially synthesizable, or not synthesizable, see the *Quick Reference for Verilog HDL* (Automata Publishing Company).

### States and Event Lists

All states associated with a Verilog model must be explicitly declared. Typically, states may be declared by using `reg` variables or instantiating storage/sequential cells from the target technology library.

Using reg variables ensures that the design can easily be remapped to any new libraries.

Constructs such as if... else, and case, can result in unnecessary registers being synthesized when all the possible states or values are not specified. If all the values are specified, the if... else construct can be used to model multiplexor logic. Thus, the constructs in Figure 7.6 result in a combinational multiplexor.

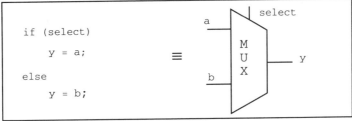

**Figure 7.6**  Multiplexor model

In contrast, the incomplete if... else construct in Figure 7.7 will result in the synthesis of unnecessary registers. This approach can be used to infer registers and preserve values across clock boundaries.

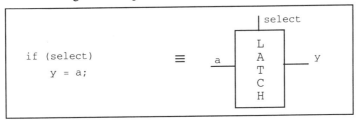

**Figure 7.7**  Set-Register model

Incomplete activation lists in an always block may also result in differences between the HDL and the gate level simulations. In the following example, the signal c is not in the list of signals entering the always block, therefore y is not re-evaluated when c changes. The simulated behavior is not that of the synthesized 3-input AND gate.

```
always @ (a or b)
    y = a & b & c;
```

The correct way to obtain the intended AND logic is

```
            always @ (a or b or c)
                y = a & b & c;
```

## Arithmetic and Relational Operators

The arithmetic operators `+`, `-`, `*`, `/` and `%` are supported by synthesis tools. Relational operators such as `<`, `>`, `>=`, `<=`, `!=` and `==` are also directly supported by synthesis tools. Figure 7.8 illustrates a synthesis model using the comparison operator. Figure 7.9 illustrates the corresponding logic schematic diagram.

```
module comparator (out, ain, bin) ;

    output out;
    input [3:0] ain, bin;

    wire [3:0] ain, bin;
    wire out;

    // comparing 2 integer values
    assign out = ain < bin ;

endmodule
```

**Figure 7.8** Relational operator `<`

## Flip-flops

State (sequential) elements can be inferred from Verilog constructs such as `@(posedge clock)`. Consider the example of a 4-bit adder, illustrated in Figure 7.10. The `@(posedge clock)` construct in Figure 7.11 is used by the synthesis tool to infer flip-flops. In the example, data is latched in at one edge and the outputs are latched out at the next edge. State assignments are inferred by the synthesis tool from the continous assignment statement, `assign`.

We can achieve the same 4-bit adder functionality by referencing a component from the ASIC library. However, doing so will restrict implementation of the design to the particular target library.

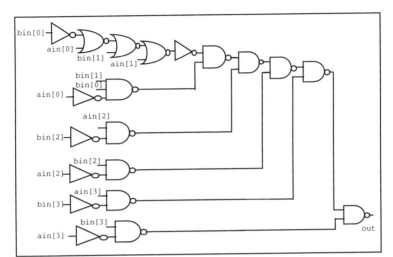

**Figure 7.9** Schematic of model using < operator

**Delays**

Delay and timing information contained in a Verilog description may be prohibited, or ignored by the synthesis tool because they imply a dependence on technology. Thus the construct:

```
always #50 clock = ~clock;
```

is accepted by simulation tools, but the delay is either ignored or not permitted by synthesis tools. RTL assignments in Verilog using the non-blocking assignment construct, <= are supported by synthesis tools, but the delay in the assignments is ignored. Thus the statement

```
a <= #delay a + b ;
```

may result in synthesized logic that does not agree with the RTL simulation, if the clock cycle is less than the delay. Synthesis tools ignore the delay and performs the RTL assignment only.

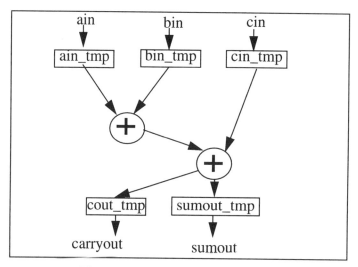

**Figure 7.10** 4-bit adder — dataflow

```
module fourbitadder(sumout, carryout, ain,
            bin, cin, clock);

    output [3:0] sumout;
    output carryout;
    input [3:0] ain, bin;
    input cin, clock;
    wire [3:0] ain, bin, sumout_tmp;
    wire cin, carryout_tmp;
    reg [3:0] sumout, ain_tmp, bin_tmp;
    reg carryout, cin_tmp;

    always @(posedge clock) begin
        carryout = carryout_tmp;
        sumout = sumout_tmp;
        cin_tmp = cin;
        ain_tmp = ain;
        bin_tmp = bin;
    end

    assign {carryout_tmp,sumout_tmp} =
            ain_tmp + bin_tmp + cin_tmp;

endmodule
```

**Figure 7.11** 4-bit adder — implementation

Timing is not inferred from an HDL description. A set of run-time constraints is used to specify the desired timing characteristics for the design and the synthesizer uses this information to evaluate timing correctness. Synthesis tools permit timing constraints to be specified on data paths and permit various area versus speed trade-offs.

**Event Control**

Some types of event control constructs may be prohibited in our HDL description. The use of Verilog declarations and statements such as `initial, forever` and `while` may be restricted or not allowed at all. These constructs are still widely used outside synthesized modules, for example, to generate test vectors or model external interfaces. In Figure 7.12, for the module `testbench`, the delay parameter cycle and the test vectors are not synthesizable and the module is used only as a test environment for verifying the synthesizable adder in Figure 7.11.

```
module testbench;
    wire [3:0] sumout;
    wire carryout;
    reg [3:0] ain, bin;
    reg cin, clock;
    integer i, j; parameter cycle = 100;
    fourbitadder INST(sumout, carryout, ain,
         bin, cin, clock);
    // adder4 INST(sumout, carryout, ain, bin,
    // cin, clock);
    initial clock = 0; // non-synthesizable clock
    always #(cycle/2) clock = ~clock; // generator

    always @(posedge clock) begin
        cin = 0; ain = 0; bin = 0;
        for ( i = 0; i <= 15; i = i + 1) begin
            #cycle ain = i;
            for ( j = 0; j <= 15; j = j + 1)
                #cycle bin = j;
        end
        #cycle cin = 1;
        for ( i = 0; i <= 15; i = i + 1) begin
            #cycle ain = i;
            for ( j = 0; j <= 15; j = j + 1)
                #cycle bin = j;
        end                              continued
```

**Figure 7.12** Stimulus for 4-bit adder

```
                #cycle $finish;
            end
            initial begin
             $monitor("%0d ",$time,, "clock = ", clock,
                    " cin = ", cin,
                    " ain[0] = ", ain[0],
                    " ain[1] = ", ain[1],
                    " ain[2] = ", aib[2],
                    " ain[3] = ", ain[3],
                    " bin[0] = ", bin[0],
                    " bin[1] = ", bin[1],
                    " bin[2] = ", bin[2],
                    " bin[3] = ", bin[3],
                    " s[0] = ", sumout[0],
                    " s[1] = ", sumout[1],
                    " s[2] = ", sumout[2],
                    " s[3] = ", sumout[3]);
            end
        endmodule
```

**Figure 7.12** Stimulus for 4-bit adder (continued)

## Unknown Values and High Impedances

The explicit X (unknown) value can be used to express error conditions during simulation of a Verilog model, but the X value may not be allowed in an assignment that is to be synthesized. The example in Figure 7.13 generates an X if an illegal combination of arithmetic flags is set.

```
'define zero 2'b00
'define positive 2'b01
'define negative 2'b10
'define error 2'bxx
 function [1:0] catch_bad_arith_flags ;
    input zero_flag, negative_flag;
    begin
      catch_bad_arith_flags = zero_flag ?
      (negative_flag ? 'error : 'zero):
      (negative_flag ? 'positive :'negative)
    end
 endfunction // catch_bad_arith_flags
```

**Figure 7.13** Unknown assignment

## Verilog HDL for Synthesis

Synthesis tools may interpret an x as a don't care condition and make use of this information in logic minimization. Thus the example in Figure 7.13 may be accepted by a synthesis tool, but the resulting gate level implementation will not generate an x on its output. Since there is no X in an actual logic, a gate level design cannot make a test for this value.

The casex construct can be used to simplify complicated states containing a number of don't care states. This can be useful for describing don't care conditions, as in the casex example illustrated in Figure 7.14.

```
'define state1 3'b000
'define state2 3'b001
'define state3 3'b01x
'define state4 3'b100
'define state5 3'b1xx

reg [2:0] state

 casex (state)
       'state1: out = 0;
       'state2: out = 1;
       'state3: out = 2;
       'state4: out = 3;
       'state5: out = 4;
  endcase
```

**Figure 7.14** casex assignment

```
'define zero 1'b0
'define one 1'b1
'define highz 1'bz
'define unknown 1'bx
 function unknown_found ;
    input to_test;
    begin case (to_test)
                'zero : unknown_found = 0;
                'one :  unknown_found = 0;
                'highz : unknown_found = 0;
                'unknown : unknown_found = 1;
         endcase
     end
  endfunction
```

**Figure 7.15** Simulation x-detect

However, incomplete descriptions utilizing X states may also generate undesired random logic. Logic synthesis tools would not generate the expected test for x in the example of Figure 7.15.

This example is simple, but there are other situations in which explicit use of unknowns is less obvious. Tests for unknown states are sometimes used in Verilog to help initialize state machines, as shown in the example of Figure 7.16.

```
'define zero 2'b00
'define positive 2'b01
'define negative 2'b10
 module foo2 (cs, in1, in2, ns)
     input [1:0] cs;
     input in1, in2;
     output [1:0] ns;

     function [1:0] generate_next_state;
         input[1:0] current_state;
         input input1, input2;
         reg [1:0] next_state;

         // input1 causes 0->1 transition
         // input2 causes 1->2 transition
         // 2->0 illegal and unknown states go to 0
     begin
       case (current_state)
         'zero :next_state = input1 ? 2'h1 : 2'h0;
         'postive : next_state = input2 ?
                         2'h2 : 2'h1;
         'negative : next_state = 2'h0;
         'default: next_state = 2'h0;
       endcase
      generate_next_state = next_state;
     end

  endfunction // generate_next_state

       assign ns = generate_next_state (cs,
                         in1,in2);
endmodule
```

**Figure 7.16** Comparison to x

In this example, if the current state is unknown, the next_state will be 2'h0. This is because the default construct

210

will match the unknown state and cause the state assignment to occur. Although the example in Figure 7.16 might be accepted by a synthesis tool, the resulting gate level design cannot perform the same function. X or `unknown` does not occur in real gates. As a result, no gate can detect the state value `unknown` and cause a state transition. If a gate level design (or the physical system) were to initialize to a 1 state, it would require an input to be asserted before a transition to the 0 state.

Similarly, comparison to Z is not permitted or is ignored by a synthesis tool. Thus in the `casez` statement, the z is ignored. Assignments to Z may be permitted and may be used to tri-state a variable. Thus assignments such as:

```
output = 3'bzzz;
```

may be used to assign high impedance to the output bus.

Following these restrictions means that we have to forego some of the higher level Verilog modeling constructs and use only the RTL constructs. We can choose to begin with a high level Verilog description and then refine it to meet synthesis restrictions during decomposition into modules. However, in practice most designers prefer to write their Verilog descriptions for synthesis from the start. Partitioning the simulation environment and the RTL synthesizable blocks into separate modules is a useful modeling approach. Future synthesis tools may be able to handle more abstract Verilog descriptions.

## Implementation Examples

There are a number of strategies for synthesizing and optimizing a large Verilog design. This section discusses some of the issues in this process, using two different examples. The first example is a traffic light controller, which illustrates how finite state machines can be synthesized from Verilog descriptions. The second example is a Verilog model for the AMD2910 microcontroller. These designs show the different types of modules for which synthesis approaches are likely to be used.

## Traffic Light Controller

The traffic light controller (Figure 7.17) is a single-way traffic signal generator that uses a finite state machine to implement a sequence of red, yellow and green signals based on the value of a counter signal, count. The Verilog model for the traffic light controller is partitioned into three modules.

1. The timer module is the top-level block and instantiates the state and counter modules.
2. The counter module is a 5-bit counter.
3. The state module describes a finite state machine (FSM) representing the traffic light states.

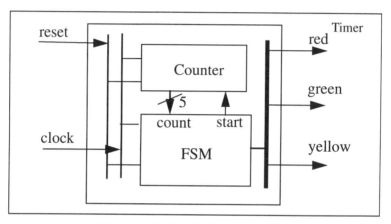

**Figure 7.17** Traffic light controller

Figure 7.18 shows the top-level structural model of the traffic light controller. The timer module has two parts:

1. Declarations of ports and signals
2. Instantiation of the state machine and the counter.

**Counter Functional Model**

The counter module shown in Figure 7.19, consists of a reset and count operation. The reset operation is initiated on the active edge of the reset signal and it resets the counter to zero. If the reset is not active, the counter continues to count and sets the output of the counter to the count_N signal. The counter generates a 5-bit count signal on the start of the count signal. On the start signal, the counter is initialized to one, and it continues to increment the count value on the posedge of the clock.

```
module timer (green, yellow, red, clk,
        reset);
    // port declarations
    output green;
    output yellow;
    output red;
    input clk;
    input reset;
    wire [5:1] count;
    wire start;

    state state (start, green, yellow,
            red, clk, reset, count);
    counter counter (count, start, clk,
                reset);
endmodule
```

**Figure 7.18** Timer — top-level module

### Finite State Machine (FSM) Functional Model

The functional model shown in Figure 7.21 describes the signal declarations, the state machine block, and its initialization block. The states, green, yellow, and red are implemented using a finite state machine.

### FSM Initialization

The FSM is reset to zero if the `reset` signal makes a posedge transition. If the reset signal is not active, the state machine implements the finite state machine describing the output states. The reset and initialization operation is shown in Figure 7.19.

### FSM Implementation

The FSM is implemented using a `casex` statement. Combinational logic is used to initialize the states and a `casex` operator is used to describe the states and the transition between the states. The state diagram in Figure 7.22 and the Verilog model in Figure 7.23 have a one to one correspondence.

```
module counter (count, start, clk, reset);
    input clk;
    input reset, start;
    output [5:1] count;
    wire clk;
    reg [5:1] count_N;
    reg [5:1] count;

    // reset operation
    always @ (posedge clk or posedge reset)
        begin : counter_S
        if (reset) begin
            count = 0;
            // reset logic for the block
        end
        else begin
            count = count_N;
        end
    end

    // count operation
    always @ (count or start)
        begin : counter_C
        count_N = count;
        // initialize outputs of the block
        if (start) count_N = 1;
        else count_N = count + 1;
    end
endmodule
```

**Figure 7.19**  Counter model

```
always @ (posedge clk or posedge reset)
    begin : fsm_S
    if (reset) begin
        fsm = 0;
    end
    else begin
        fsm = fsm_N;
    end
end
```

**Figure 7.20**  Finite state machine initialization

# Verilog HDL for Synthesis

```
module state (start, green, yellow,
              red, clk, reset, count);

    output green, yellow, red;
    output start;
    input clk;

    input reset;
    input [5:1] count;

    reg green;
    reg yellow;
    reg red;
    reg start;
    reg [5:1] count_N;
    reg [2:1] fsm_N;
    reg [2:1] fsm;

    // fsm-initialization

    // fsm-implementaion

endmodule
```

**Figure 7.21** Finite state machine model

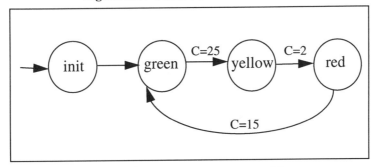

**Figure 7.22** Traffic light state machine

The FSM block implementation is shown in Figure 7.23. All the states are explicitly described and the default sets the states to 'bx and is used by the synthesis tool for optimization.

215

```
always @ (fsm or count)
    begin : fsm_C
    parameter [2:1] init = 0, g = 1,
                    y = 2, r = 3;
    red = 0;
    fsm_N = fsm;
    green = 0;
    start = 0;
    yellow = 0;

    casex (fsm)
        init: begin
            start = 1;
            fsm_N = g;
        end
        g: begin
            green = 1;
            if (count == 25) begin
                start = 1;
                fsm_N = y;
            end
        end
        y: begin
            yellow = 1;
            if (count == 2) begin
                start = 1;
                fsm_N = r;
            end
        end
        r: begin
            red = 1;
            if (count == 15) begin
                start = 1;
                fsm_N = g;
            end
        end

        default: begin
            red = 'bx;
            fsm_N = 'bx;
            green = 'bx;
            start = 'bx;
            yellow = 'bx;
        end
    endcase
```

**Figure 7.23** FSM implementation

## AMD2910 Microcontroller

The AMD2910 is a 12-bit microcontroller that is used as an address sequencer for controlling the sequence of execution of microcode instructions, stored in read-only memory. External addresses of up to 4K words are permitted by the 12-bit internal data path. The device architecture (from the AMD data sheet, January 1989) is illustrated in Figure 7.24.

**Partitioning the Microcontroller**

Functional units as specified in the AMD data sheets are selected as the major functional blocks to implement the microcontroller. Conscious use and reuse of functional blocks ensure that the intended architecture is achieved with minimum overlap. In subsequent phases of the design process, each constituent module is implemented. Design teams may partition the blocks across team members along the lines of the functional blocks identified in Figure 7.24.

We will provide a brief description of each constituent module of the microcontroller; however, we will not cover detailed description of the chip. As each RTL description is completed, simulation is used to verify the functional behavior of the description. Simulation environments using test drivers is written for each module. These drivers instantiate the RTL module under test. Input stimulus is used to drive the input ports and output results are compared with expected results.

The Verilog model of the AMD2910 chip is decomposed into seven modules. Partitioning is shown using dotted boxes in Figure 7.24.

1. The `register` module is a 12-bit register/counter used for setting up loops and has a ZERO detect signal.

2. The `stack` module is a 9-word deep stack, with an overflow detect signal FULL_.

3. The `incrementor` module is a 12-bit incrementor and program register, and is used to count the microprogram.

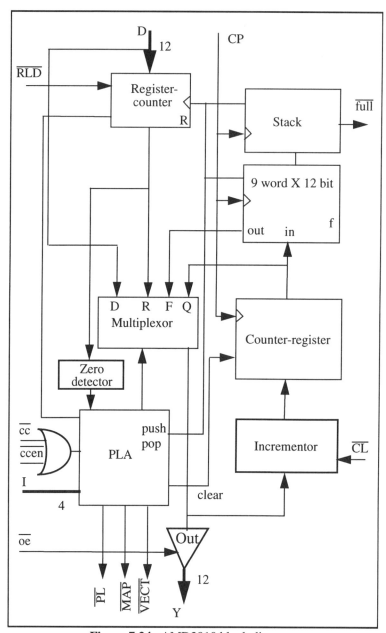

**Figure 7.24** AMD2910 block diagram

4. The `mux` module is a 12-bit 4-input data multiplexor that allows the selection of the next address from the output of the stack, the register/counter, the incrementor and an external address.

5. The `pla` module contains the instruction decoder and performs the control logic for the design.

6. The `out` module is a 12-bit tri-state output block.

7. The `M2910` is the top level functional block for the chip that instantiates all of the above functional units.

**Register Model**

The register module in Figure 7.25 contains state elements and is controlled by the input `register_cnt`. A `ZERO` detect signal is computed using bit-wise operation. Depending on the `register_cnt` input, the output is decremented or the input is written directly to the output.

The state elements represented by the `reg` variables are synthesized as edge-triggered flip-flops. The simple functional operations such as the 1-bit equality test are mapped into random logic by the synthesis tool. The "-" operator can considerably complicate the synthesis of this module.

Some tools require a designer to guide the implementation of arithmetic functions, for example, it may be necessary to express the "-" operation in more complex terms — such as explicit carry-look-ahead adder form — in order to meet timing constraints.

**Stack Model**

The `stack` module defines several registers which are set within a `case` statement as shown in Figure 7.26. The model is a `LIFO` (last-in first-out) stack network with `PUSH`, `POP`, `HOLD` and `CLEAR` instructions. The stack holds nine of the 12-bit input addresses and has an overflow signal `FULL_`.

The handling of the `+` and `>` operators in this module is similar to the `-` operator. Some synthesis tools may require a more elaborated form of the operation in order to produce timing-efficient gates.

```
`define REGDATA 2'b00
`define REGONEBIT 2'b01

module register (R, ZERO, register_cnt, D,
             RLD_, CP);
    output [11:0] R;
    output ZERO;
    input [11:0] D;
    input [1:0] register_cnt;
    input    RLD_, CP;
```

**Figure 7.25** Register module...

```
    reg [11:0] R_IN, R ;
    reg ZERO;

    always @ ( register_cnt or D or RLD_ )
         begin if ( RLD_ == 1'b0 )
               R_IN = D ;
         case ( register_cnt )
          `REGDATA :
               R_IN = D ;
          `REGONEBIT :
               R_IN = R_IN - 12'b1 ;
          default :
               R_IN = 12'bx ;
         endcase
    end

    always @ (posedge CP) begin
         R = R_IN ;
    end

    always @ ( R ) begin
         assign ZERO = ~(|(R));
    end
endmodule
```

**Figure 7.25** Register module (continued)

```verilog
`define HOLD 2b'00
`define POP 2b'01
`define PUSH 2b'10
`define CLEAR 2b'11
 module stack (F, FULL_, stack_cnt, UPC, CP);
    output [11:0] F;
    output FULL_;
    input [1:0] stack_cnt;
    input [11:0]UPC;
    input CP;
    reg [11:0] F;
    reg [8:0] stack_BIT0, stack_BIT1,..........,
              stack_BIT10, stack_BIT11;
    reg FULL_;
    reg [3:0] ADDR;

    always @ (posedge CP)
     case (stack_cnt)
    // may use special information such as
    // parallel case
         `HOLD :        ;
         `POP  :
                if (ADDR != 0) ADDR = ADDR - 1;
         `PUSH : begin
                ADDR = ADDR + 1;
                stack_BIT0[ADDR] = UPC[0];
                stack_BIT1[ADDR] = UPC[1];
                .......................
                .......................
                stack_BIT10[ADDR] = UPC[10];
                stack_BIT11[ADDR] = UPC[11];
             end
         `CLEAR :ADDR = 0 ; // CLEAR
    endcase
    if ( ADDR >= 9 ) begin
                FULL_ = 1'b0;
                ADDR = 9 ;
                end
    else FULL_ = 1'b1;
                F[0] = stack_BIT0[ADDR];
                .......................
                F[10] = stack_BIT10[ADDR];
                F[11] = stack_BIT11[ADDR];
     end
endmodule
```

**Figure 7.26** Stack module

**Incrementor Model**

The `incrementor` module in Figure 7.27 contains a 12-bit `incrementor` and a 12-bit register. As with the `stack` module, a `case` statement is used to increment or decrement the module. The module also uses the `+` operator. It is interesting to note that the `+` operation in this module is actually used for an increment operation, and not a full addition. A synthesis tool should be able to recognize this situation, and build an incrementor instead of the more complex adder. The remaining four modules contain only combinational logic.

```verilog
`define UPCINC 1'b0
`define UPCDEC 1'b1

module incrementor (UPC, incrementor_cnt,
            Y_IN, CI, CP);

    output [11:0] UPC;
    input [11:0] Y_IN;
    input incrementor_cnt;
    input CI, CP;

    reg [11:0] UPC;
    reg [11:0] UPC_IN;

    always @ (Y_IN or incrementor_cnt or CI)
        begin
          case ( incrementor_cnt )
              `UPCINC :
                    UPC_IN = Y_IN + CI;

              `UPCDEC :
                    UPC_IN = 0;

              default :
                    UPC_IN = 0 ;
          endcase
    end

    always @ ( posedge CP ) begin
        UPC = UPC_IN ;
    end

endmodule
```

**Figure 7.27**  Incrementor module

## Multiplexor Model

The mux module is a simple four-way multiplexor as shown in Figure 7.28.

```
'define DATA 2'b00
'define REG 2'b01
'define STACKIN 2'b10
'define UPCOUT 2'b11

module mux (Y_IN, mux_cnt, D, R, F, UPC);

    output [11:0] Y_IN;
    input [1:0] mux_cnt;
    input [11:0] D, F, R, UPC;

    reg [11:0] Y_IN;

    always @ ( mux_cnt or D or R or F or
               UPC ) begin
      case ( mux_cnt )
            'DATA :
                 Y_IN = D ;
            'REG :
                 Y_IN = R ;
            'STACKIN :
                 Y_IN = F ;
            'UPCOUT :
                 Y_IN = UPC;
      endcase
    end
endmodule
```

**Figure 7.28** Multiplexor module

This module looks simple since it only requires the synthesis tool to map a four-way case statement to a four-way multiplexor. However, the semantics of the Verilog case statement implies that a multiplexor may be used only if all possible cases are enumerated. If all states are not enumerated, some tools produce a priority encoder implementation. Some synthesis tools need a directive to resolve this situation and specify that a multiplexor is acceptable. The directive may be in the form of Verilog comments (//) or compiler directives ('ifdef SYNT).

## PLA Model

The `pla` module (illustrated in Figure 7.29) is the most complex, using the micro instruction and status information to control the actions of the counters, multiplexor, and output modules. Since the controller does not store any information it contains only combinational logic. The output depends on the input instruction and some condition signals (CC_, CCEN_).

```verilog
`define JZ 4'b0000
`define CJS 4'b0001
// ..... Micro Instructions
`define CONT 4'b1110
`define TWB 4'b1111

`define INCNT 1'b0
`define INCHI 1'b1

`define STKCNT 2'b00
`define DATCNT 2'b01
`define REGCNT 2'b10
`define MUXCNT 2'b11

module pla (PL_, MAP_, VECT_, register_cnt,
        mux_cnt, stack_cnt,
        incrementor_cnt,
        ZERO, I, CC_, CCEN_);

    // input-output-reg declaration

    always @ ( CC_ or CCEN_ ) begin
        PASS = ~CC_ | CCEN_ ;
    end

    // microcode instruction-operation
    always @ ( I or ZERO or PASS ) begin
        stack_cnt = 'STKCNT;// NC
        mux_cnt = 'MUXCNT;// UPC
        register_cnt = 'REGCNT://  NC
        incrementor_cnt = 'INCNT ;
        PL_ = 'INCNT ;// LOW
```
**continued**

**Figure 7.29** PLA module

```verilog
             MAP_ = 'INCHI ;
             VECT_ = 'INCHI ;

             // micro-code --> operation
             case ( I )
             'JZ :begin
              stack_cnt = 'MUXCNT;// CLEAR
              incrementor_cnt = 'INCHI ; // CLEAR
             end
             'CJS :begin
               if ( PASS ) begin
                    stack_cnt = 'REGCNT;// PUSH
                    mux_cnt = 'STKCNT;// D
               end
             end
             'JMAP :begin
                    // operation........
             end
             'CJP :begin
                    ..................
             'PUSHIN :begin
                         ..................
             'JSRP :begin
              stack_cnt = 'REGCNT;// PUSH

              if ( PASS )
                    mux_cnt = 'STKCNT;// D
              else
                    mux_cnt = 'DATCNT;// R
              end
             'CJV :begin
                    ..................
             'JRP :begin
                    ..................
             'RFCT :begin
                    ..................
             'RPCT :begin
                    ..................
             'CRTN :begin
                    ..................
             'CJPP :begin
                    ..................
             'LDCT :begin
                    ..................
                                              continued
```

**Figure 7.29** PLA module (continued)

```
                'LOOP :begin
                     .................
                'CONT :begin
                     .................
                'TWB :begin
                if ( PASS & ZERO ) begin
                        stack_cnt = 'DATCNT;// POP
                end
                else if ( PASS & ~ZERO ) begin
                        stack_cnt = 'DATCNT;// POP
                        register_cnt = 'DATCNT;// DEC
                end
                else if ( ~PASS & ZERO ) begin
                        stack_cnt = 'DATCNT;// POP
                        mux_cnt = 'STKCNT;// D
                end
                else if ( ~PASS & ~ZERO ) begin
                        mux_cnt = 'REGCNT;// F
                        register_cnt = 'DATCNT;// DEC
                end
                end
                endcase
        end
endmodule
```

**Figure 7.29** PLA module (continued)

It is likely that the functions performed by the PLA module are part of the critical timing path for an implementation of the AMD2910 design. Therefore, it is also likely that synthesis with a high degree of speed optimization provides the best solution.

**Tri-state Out Model**

The out module presents the special case of tri-state gates as shown in Figure 7.30.

This block uses a tri-state cell from the ASIC library. The TRISTATE module may also be built using the high-impedance operator as illustrated in Figure 7.31.

```
module out (Y, Y_IN, OE_);
    output [11:0] Y;
    input [11:0] Y_IN;
    input  OE_;
    /* the "tristate" module represents an
    ASIC vendor library tri-state output
    cell, with the parameters (tristated_
    data_output, data_input, tristate_en
            able_input) */
    tristate TRI11 (Y_IN[11], OE_, Y[11]),
             TRI10 (Y_IN[10], OE_, Y[10]),
             TRI9 (Y_IN[9], OE_, Y[9]),
             TRI8 (Y_IN[8], OE_, Y[8]),
             TRI7 (Y_IN[7], OE_, Y[7]),
             TRI6 (Y_IN[6], OE_, Y[6]),
             TRI5 (Y_IN[5], OE_, Y[5]),
             TRI4 (Y_IN[4], OE_, Y[4]),
             TRI3 (Y_IN[3], OE_, Y[3]),
             TRI2 (Y_IN[2], OE_, Y[2]),
             TRI1 (Y_IN[1], OE_, Y[1]),
             TRI0 (Y_IN[0], OE_, Y[0]);

endmodule
```

**Figure 7.30** Tri-state instantiation: out module

```
module TRISTATE (Y_IN, OE_, Y);
    output Y;
    input Y_IN, OE_;
    reg Y;

    always @ (OE_) begin
     if (OE_)
            Y = Y_IN;
     else
            Y = 1'bz; // assign high-impedance
    end
endmodule
```

**Figure 7.31** Tri-state module

The assignment of high impedance was used to infer the tri-state logic, but some optimization methods cannot handle tri-state signals directly. Hence some synthesis tools may need special directives to accept this module as it is written. Other tools may not be able to accept the Verilog z value at all. In this case, it is usually necessary to explicitly instantiate the desired tri-state output cell from the target ASIC library. However, this results in an HDL description that cannot be easily remapped to a new technology.

**AMD2910 Structural Model**

Finally, Figure 7.32 illustrates the M2910 top-level structural module, which instantiates all of the other submodules, but contains no functionality of its own. Since it has no native functionality, logic synthesis of this module by itself would do little more than map the description to the target netlist. However, synthesizing this module in the context of the rest of the design can be useful for further system level verification.

```
module AMD2910 (VECT_, MAP_, PL_, FULL_, Y,
        CP, OE_, RLD_, CI,CCEN_, CC_, I, D);

    // input-output-wire declaration
    output VECT_, MAP_, PL_, FULL_;
    output [11:0] Y;
    input  CP, OE_, RLD_, CI, CCEN_, CC_;
    input [3:0] I;
    input [11:0] D;
    wire ZERO, LOAD, DECR, HOLD, CLEAR, POP, PUSH;
    wire [11:0] R, Y_IN, F, UPC;
    wire [1:0] register_cnt, stack_cnt, mux_cnt;
    wire incrementor_cnt;

    register register (R, ZERO, register_cnt,
                    D, RLD_, CP);
    stack stack (F, FULL_, stack_cnt, UPC, CP);
    incrementor incrementor (UPC,
              incrementor_cnt,Y_IN, CI, CP);
    mux mux (Y_IN, mux_cnt, D, R, F, UPC);
    pla pla (PL_, MAP_, VECT_, register_cnt,
          mux_cnt,stack_cnt,incrementor_cnt,
          ZERO, I, CC_, CCEN_);
    out out (Y , Y_IN, OE_);

endmodule
```

**Figure 7.32** AMD2910 structural model

**Synthesizing AMD2910**

Synthesis at the top level of a design allows timing between the submodules to be refined based on the implementation of the modules. Some changes in the submodules might occur during this phase to improve the timing between the modules. In this example, the synthesis tool might increase the drive on the `Y_IN` output of the `mux` module to handle the loads from both the `incrementor` and `out` modules.

On the other hand, some global optimization could also result from this top-level synthesis. The tool might eliminate drive gates that are not necessary after evaluation of the inter-module timing. If the designer opts to "smash" or repartition the hierarchy and allow the synthesis process to swap gates among the submodules, even greater optimization may happen. This change in the hierarchy may considerably complicate the design flow, however.

Some synthesis tools may be able to handle design hierarchy directly, but others require explicit commands for each module. In such cases, the following approach provides a good solution:

1. Read in the entire design and evaluate the timing from the top-level module. This provides the timing definitions for the interfaces on all the sub-modules.

2. Synthesize each sub-module independently, using timing constraints derived from the analysis above. This allows careful tuning of the optimization constraints for each piece of the design.

3. Run optimization on the entire design to provide global optimization and correct for any changes in the submodule timing.

In some cases the designer may choose to retain some blocks which are synthesized separately. Synthesis tools provide constraints to prevent optimization of these blocks.

The optimization for power consumption can be essential for low-power designs, such as portable or battery-operated products. It is also important for designs using high-speed technologies in which using the fastest versions of all gates may result in a chip that cannot be cooled adequately.

## Design Implementation and Management

Producing a gate level design from a synthesized Verilog module is theoretically simple. A first pass gate level design can typically be accomplished very quickly. Very few synthesis commands are needed to read each module, link the design, map to the target ASIC library, and output a gate level netlist.

As long as the ASIC vendor synthesis library has no errors, a first pass design will be logically correct. However, a design produced so easily may not meet the timing, size, and power constraints of the design. The process of optimizing a design for these parameters requires detailed design constraint specification and can take numerous iterations.

**Verilog Libraries For Synthesis**

The restrictions for synthesis may affect the way Verilog libraries are defined and used. One popular technique for large Verilog designs is to create libraries of useful functions that may be shared among many designers. These libraries may include models that have been created to simulate efficiently, but may result in modules that cannot always be synthesized. As a simple example, consider a 4-bit input to a 1-bit output multiplexor, with 2 select signals. For simulation efficiency the multiplexor may be described as a user-defined primitive as illustrated in Figure 7.33.

```
primitive MUX_4_2 (Y, D0, D1, D2, D3, S0, S1);
    input D0, D1, D2, D3, S0, S1;
    output Y;
    table // D0 D1 D2 D3 S0 S1 : Y
    0 ? ? ? 0 0 : 0 ;
    1 ? ? ? 0 0 : 1 ;
    ? 0 ? ? 1 0 : 0 ;
    ? 1 ? ? 1 0 : 1 ;
    //.............
    // complete model in listing
    1 ? 1 ? 0 ? : 1 ;
    ? 0 ? 0 1 ? : 0 ;
    ? 1 ? 1 1 ? : 1 ;
    endtable
endprimitive
```

**Figure 7.33** 4-1 Multiplexor UDP model

# Verilog HDL for Synthesis

The Verilog `primitive` construct is not currently supported by some synthesis tools. For details, refer to the Quick Reference for Verilog HDL Guide. Therefore, an alternative synthesizable definition for this module might look as in Figure 7.34.

```
module MUX_4_2 (Y, D, S);
    input [0:3] D;
    input [0:1] S;
    output Y;
    reg Y;

    always @(select or D) begin
      case (select)
            0: Y = D[0];
            1: Y = D[1];
            2: Y = D[2];
            3: Y = D[3];
            4: Y = D[4];
      endcase
    end
endmodule
```

**Figure 7.34** 4-1 Multiplexor RTL model

The notion of Verilog libraries takes a new dimension during the synthesis process. Instead of simply including a presynthesized Verilog library element, a designer can synthesize the element according to its use. This approach synthesizes generic Verilog library descriptions into unique gate level implementations for each usage. Thus, the library element becomes a "soft macro" that the designer may customize using optimization during the synthesis process.

The synthesis process also requires model libraries for the ASIC vendor technology. Currently this library format varies from synthesis vendor to vendor. However, in general, synthesis libraries contain the following information:

1. Cell descriptions for every macro in the ASIC vendor library. This contains rise-time, fall-time, intrinsic timing information, pin information, capacitance, signal types (clock, test for synthesis signals), logic functionality and delay models.

2. Process parameters and area/power information for the cells.

3. Wire-load models, operating conditions, and scaling factors based on the operating conditions.
4. Default attributes for various macros.

Besides the technology library there may also be a separate symbol library. The symbol library is used for graphical symbol/schematic generation from the synthesized gate level netlist. While these libraries are required for synthesis and schematic generation, separate gate level models are required for simulation.

There are situations in which more efficient simulation makes separate simulation and synthesis libraries useful. However, such an approach adds to the design complexity since different libraries must be read in for simulation and synthesis runs. Care must be taken to ensure that the libraries are consistent so that the synthesized logic will match the original Verilog description.

**Design Planning**

When writing Verilog descriptions for a design to be implemented into gates we are faced with many decisions about the decomposition of the design into modules. Synthesis may influence this design planning and partitioning process.

By proper planning we can exercise great influence over the efficiency of the final gate level realization of the design. For example, consider a 16-bit signal COUNTER[15:0], which comes from a counter that counts up from zero. If we want to generate an ENABLE signal when the COUNTER reaches 256, we can use any of the following conditional assignments:

```
            if (COUNTER >= 256) ENABLE = 1;
                        or
            if (COUNTER == 256) ENABLE = 1;
```

However, it is more economical to implement this by checking for the eighth bit of COUNTER, i.e.,

```
            if (COUNTER[8] == 1) ENABLE = 1;
```

The first case may result in NAND/NOR gates for the entire bus. The second representation will result in either no additional gates, because `ENABLE = COUNTER[8]`, or may result in NAND/NOR gates for only one of the bits of the bus. This approach, however, may result in design descriptions that cannot be easily scaled. Thus, in the above example, if the second representation is used, it is more difficult to scale the design to generate an ENABLE signal when the COUNTER reaches 512.

Similar planning has to be put into organizing Verilog descriptions for sharing operators and building only the required number of registers. Every variable with an assigned value in an `always @(posedge)` or `always @(negedge)` block results in flip-flop implementations. We will illustrate it with two different descriptions of a 4-bit to 16-bit decoder.

```verilog
module d4x16 (inp, clk, reset, out,enand,
              enor, enxor);
   input [0:3] inp;
   input reset, clk;
   output [0:15] out;
   output enand, enor, enxor ;
   reg [0:15] out;
   reg enand, enor, enxor ;

   always @ (posedge clk) begin
         if (reset) begin
               out = 0 ;
               enand = 0 ;
               enor = 0 ;
               enxor = 0 ;
         end
         else begin
               out = 0;
               out[inp] = 1 ; // simpler modeling!
               // or out = 16'b8000 >> inp ;
               enand = &(out);
               enor = !(out);
               enxor = ^(out);
         end
   end

endmodule
```

**Figure 7.35** Inefficient 4-bit to 16-bit decoder

In Figure 7.35, the `reg` data type is used to hold the values of `out` and the signals `enand, enor,` and `enxor`. Since `out` is already registered, it is not necessary to register `enand, enor` and `enxor`. Registering `enand, enor,` and `enxor` will result in extra hardware without providing any extra functionality. As a result, assignments that are clocked are put in a separate always block from the other assignments (Figure 7.36). The same model may also use the `case` construct.

```verilog
module d4x16 (inp, clk, reset, out,
              enand, enor, enxor);
    input [0:3] inp;
    input reset, clk;
    output [0:15] out;
    output enand, enor, enxor ;
    reg [0:15] out;
    wire enand, enor, enxor ;

    always @ (posedge clk) begin
        if (reset)
            out = 16'b0 ;
        else begin
            out = 0;
            case (inp) // directives to tool may
                0: out[0] = 1; // be used as comments
                1: out[1] = 1;
                // .....
                // .....
                15: out[15] = 1;
        end
    end

    assign enand = &(out);
    assign enor  = |(out);
    assign enxor = ^(out);

endmodule
```

**Figure 7.36** Efficient 4-bit to 16-bit decoder

**Design Partitioning**

The design partitioning that we specify using the HDL design hierarchy plays an important role in the final design implementation. The determination of module boundaries in a hardware description is based on several factors:

- Ease of understanding
- Physical design considerations
- Project management issues
- Design structure
- Module size considerations

**Ease of Understanding**

Ease of understanding is the most important criterion for a good design. It is a good practice to structure designs so that each low-level module has a self-contained, well-defined function. Higher-level modules should be grouped together into logically related modules.

As an example, consider the ALU illustrated in Figure 7.37. The ALU performs addition/subtraction, multiplication, division, and logic operations. One reasonable organization is to define each of these four functions in a separate Verilog module. A higher-level module to represent the entire ALU can instantiate the four arithmetic modules. As a contrast, grouping pairs of functions into arbitrary modules and then joining these at a yet higher level would detract from the clarity of the design.

**Physical Design Consideration**

Physical design considerations may cause modules to be divided in ways that are not optimal for understanding. It is generally a good idea to write Verilog descriptions in which chip and board boundaries correspond to module interfaces. Other hard physical design constraints may include die-size, package and pin limitations. This partition at the physical level makes synthesis and test vector capture much easier. In the above example, the logic and add/subtract operations might be performed by one chip, while another chip handles the multiply and divide operation. In such a case, it is desirable to have modules that correspond to the physical partitioning.

**Project Management Issues**

Project management issues can also affect the manner in which a Verilog design is broken into modules. A single function split between two engineers is handled most effectively when each has a distinct

Verilog module to write, debug, and control. Managing design complexity and splitting the implementation between various engineers may affect the design partitioning. The effects of file input/output during simulation may also affect the partitioning of the blocks. For example, a case statement provides a good mechanism for expressing a multi-way decision and could be used to model the ALU efficiently in one module. This is illustrated in the simple ALU module described in Figure 7.37.

```verilog
`define OPCODE0 3'h0
`define OPCODE1 3'h1
`define OPCODE2 3'h2
`define OPCODE3 3'h3
`define OPCODE4 3'h4
`define OPCODE5 3'h5
`define OPCODE6 3'h6
`define OPCODE7 3'h7
 module alu (alu_out,zero,opcode,
                 data,accum,clock);
     input [7:0] data,accum;
     input [2:0] opcode;
     input clock;
     output [7:0] alu_out;
     output zero;

     always @(negedge clock)
      case (opcode)
            `OPCODE0: alu_out = accum;
            `OPCODE1: alu_out = accum;
            `OPCODE2: alu_out = accum + data;
            `OPCODE3: alu_out = accum & data;
            `OPCODE4: alu_out = accum ^ data;
            `OPCODE5: alu_out = data;
            `OPCODE6: alu_out = accum;
            `OPCODE7: alu_out = accum;
       endcase

     always @(accum) begin
            if (!accum) zero = 1;
            else zero = 0;
       end
  endmodule
```

**Figure 7.37** ALU Verilog model

**Design Structure**

Logic synthesis adds yet more dimensions to the structuring of Verilog designs. In general, it is easier and more efficient to optimize within modules than across module boundaries. A design not properly partitioned can result in unsatisfactory performance by synthesis tools.

```verilog
module adder (A,B,C,D,clk,reset,out);
    output[0:7] out;
    input[0:7] A,B,C,D;
    input clk,reset;
    reg [0:1] fsm, fsm_N;
    reg[0:7] out;
    parameter state0 = 0,state1 = 1,state2 = 2;

    always @(posedge clk or posedge reset)
        begin:adder_reset // reset block
            if (reset) fsm = 0;
            else fsm = fsm_N;
        end

    always @(fsm) begin : adder_func
        out = 0;
        fsm_N = fsm;
        casex(fsm)
            state0: begin
                out = 0;
                fsm_N = state1;
            end
            state1: begin
                out = A + B;
                fsm_N = state2;
            end
            state2: begin
                out = C + D;
                fsm_N = state0;
            end
            default: begin
                out = 8'bx ;
                fsm_N = 'bx ;
            end
        endcase
    end
endmodule
```

**Figure 7.38** Adder without resource sharing

Grouping of logic in ways that provide more efficient realizations require the designer to properly analyze the design datapath. For example, the two adders shown in Figure 7.38 and Figure 7.40 provide the same functionality, but the one in Figure 7.40 results in a smaller design.

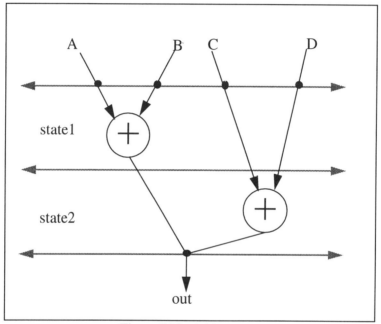

**Figure 7.39** Adder states

The second implementation uses the fact that the two additions are never used in the same state and generates a more efficient realization. The data flow diagram is illustrated in Figure 7.39. The `function` declaration is used to share the addition operator between the two states, `state1` and `state2`. Resource sharing is used by synthesis tools to share multiple operations that are not executed in the same clock cycle. However, take care to avoid combinational feedback loops. Some synthesis tools may provide this capability automatically, thereby freeing the designer from this partioning constraint.

Using two array indices may result in inefficient circuit description. In the example illustrated in Figure 7.41, two array indices are used to select between two arrays.

Figure 7.42 implements the same `counter_C`, but utilizes a single array using a temporary `index` and `count_temp` variable. This results in a more efficient implementation of the description.

```
module adder (A,B,C,D,clk,reset,out);
// input - output - reg parameter declarations
    function [0:7] outbus;
        input [0:7] IN1, IN2 ;
        begin : functionality
            outbus = IN1 + IN2;
        end
    endfunction
    // .... reset block
    always @(fsm)
    begin : adder_func
        ......
        casex(fsm)
        state0: begin
            out = 0;
            fsm_N = state1;
        end
        state1: begin
            out = outbus(A, B) ;
            fsm_N = state2;
        end
        state2: begin
            out = outbus(C, D);
            fsm_N = state0;
        end
        default: begin
            ........
        end
        endcase
    end
endmodule
```

**Figure 7.40** Adder with resource sharing

**Module Size Considerations**

Synthesis algorithms tend to break down as the logic size increases. As a result, most contemporary logic synthesis tools work best when they are optimizing medium size blocks (say, a few thousand gates at a time). For this reason, it is useful to partition designs so that no module will map to more than a few thousand gates. This is especially true for critical modules that are likely to be heavily optimized during the synthesis process. Attempting to optimize too large a module may also result in the synthesis process taking an unacceptably long time to converge to a solution.

```
always @ (select or count or index1 or
         index2)
    begin: counter_C
    if (select)
        count_1 = count[index1] ;
    else
        count_2 = count[index2] ;
end
```

**Figure 7.41** Two array indices

```
always @ (select or count or index1 or
         index2) begin : counter_C

    if (select)
        index = index1 ;
    else
        index = index2 ;
    count_temp = count[index] ;
    if (select)
        count_1 = count_temp ;
    else
        count_2 = count_temp ;
end
```

**Figure 7.42** One array index

Allocation and partitioning determine the achievable optimization and therefore it is advisable to partition the Verilog modules taking the criticality and optimization of each module into consideration.

Some synthesis tools provide the ability to group arbitrary HDL statements or clusters of gates into temporary levels of hierarchy. This provides a way to define modules during synthesis runs, and to apply different types of optimization to different portions of the same module. However, this approach is very sensitive to minor changes in the module description and complicates management of the design.

## Simulation and Verification

The synthesis process results in a gate level design that matches the functionality of the Verilog module. Gate-level simulations need to be performed because of the timing introduced at the gate level and also because of the necessity to verify the integrity of the gate level design, synthesis library and the synthesis tool. Simulation of the RTL design verifies the functionality of the design. Gate-level simulations can be used to verify both timing and functionality.

Designers may choose to run only a subset of the full test vector suite on the synthesized gates, depending upon their measure of confidence in the synthesis tools. At a minimum, all major functions should be exercised in this subset. The designer may use functional vectors to verify the RTL description and may utilize these to provide a first level of verification for the synthesized logic. Additional vectors may be needed at the gate level for fault simulation and timing analysis.

One technique used to verify manually designed gates against an HDL description can also be used for synthesized gates. If fault grading or ATPG (automatic test pattern generation) tools are available, we can identify a subset of tests which result in a high degree of fault coverage at the gate level. It may also be possible to utilize Verilog's PLI (Programming Language Interface) to write C-programs to read both ASCII and binary procedural vector formats, which may be the output of these ATPG tools.

If the results of these tests match at the RTL and the gate level, the designer may have a reasonable degree of confidence in the final design. In the absence of such tools, users have to manually generate the vectors that provide a high degree of fault coverage for the design.

Timing verification for synthesized gates is a harder problem. Even with vendor-certified ASIC models, a synthesis tool may not be able to adequately estimate all delays needed to produce a design that satisfies timing. In most cases, a designer should apply the same degree of timing analysis to synthesized gates as to manually designed gates.

There are several reasons why synthesis tools typically have a harder time meeting timing criteria than functionality:

1. It can be difficult to describe to a synthesis tool all timing requirements for a design. The use of multiple clocks, asynchronous circuits, and latches make it all the more difficult to specify the timing details.

2. Some synthesis tools have limited capability for using placement information. Especially in large chips, pre-route block placement (floorplanning) is used by ASIC vendor delay estimation tools when doing front-annotation.

   This placement is also used as a starting point for the place and route process. If a synthesis tool cannot process this information when doing its timing calculations, it may make inaccurate assumptions about delays between modules.

3. When timing analysis is run on an ASIC using post route (back annotation) delays, it is not unusual to find a few problems. Unexpectedly short or long signal routes may result in deviations from the front-annotation estimates. Some synthesis tools can rerun their delay calculations using post-route delays, and warn the designer of any timing violations.

## Summary

Logic synthesis tools in conjunction with a Verilog HDL simulator provide a powerful and flexible design environment. The design process can be summarized in six main stages as follows:

1. Describe the design using the Verilog language. Follow the synthesis restrictions for any ASIC module to be synthesized. Make use of the Verilog hierarchical module structure, tasks and functions to make the description more understandable and manageable.

2. Careful partitioning and planning of the design will result in a more cost effective realization. It is sometimes desirable to obtain a first-pass gate level description from

the algorithmic Verilog description. However, the designer needs to restructure the original design in order to achieve required performance.

3. Simulate the description using a simulator that supports Verilog. Verify the architectural and functional correctness of the design. Perform timing analysis on major module interfaces.

4. Synthesize each Verilog module to generate a gate level implementation of the design. Schematic editors and text editors may be used to generate non synthesized gate level HDL netlists.

5. Run the full set of simulations using the gate level design. Perform detailed timing analysis using actual gate delays and estimated (pre-layout) delays for ASIC and board wires.

6. Place and route the ASIC and board designs. Calculate wire delays based on actual (back annotation) route results.

7. Run final simulations and timing analysis using actual gate and wire delays.

This chapter covered issues that arise when synthesis is used to help transform a Verilog RTL and logic description into a gate level design. This process leads to several significant areas of savings in project time and complexity. It is important, however, to regard synthesis as just another tool in the design process. Its assumptions should be carefully established and its results should be thoroughly analyzed.

## Exercises

1. Using a shift register, design a circuit that takes in a serial bit stream, feeds it through a shift register, and raises the flag FOUND if a sequence of eight ones followed by 01010101 arrives. Maximize the speed performance for the FOUND flag. Develop a test module that verifies the above functionality.

2. Design a serial parity-bit generator that receives coded messages and adds a parity-bit to every m-bits message, so that the outcome is an error-detecting coded message. The

strings of m-bits are spaced apart by single time units. The parity-bits are to be inserted in the appropriate spaces, so that the resulting outcome is an error-detecting coded message. Assume that even parity is used.

3. As a variation to the above design, design a parity-bit for a binary-coded-decimal (b-c-d) messages. Each binary-bit is synchronized with a clock pulse input. Assume the circuit establishes odd-parity by generating the parity-bit.

4. A traffic light controller controlling one traffic signal was designed in the chapter. Improve the traffic signal controller to a four-way traffic light controller, such that the states change dependant on a serial queue signal. We suggest that you build a four-way controller and use the traffic light controller presented in this chapter as a starting point.

5. Rewrite models rsisc.v and rsisc_pipe.v in the "Pipelined Processor" chapter so that they can be synthesized.

6. Design an automatic cruise control circuit which performs the operations in Figure Figure 7.43. The circuit is described in further detail in "Comparing the Computer-Aided Systems in Action" (VLSI Design, November, 1983, pp. 24-35). Generate a test module to verify the above functionality.

7. Design an arithmetic module-10 function, which has a single 8-bit word as input and gives the remainder after dividing by 10.

8. Design a synthesizable Verilog model of the AMD8254 programmable interval counter/timer (AMD data sheet, January 1989). Figure 7.44 shows a block diagram of the chip. A summary of the functionality is described below, please refer to the data sheets for further documentation. The design consists of a transceiver to interface the chip to the system bus, read/write logic, a control word register and three counters. Verify using simulation the RTL and gate level design.

**Summary of AMD8254 Functionality**

**Read/Write Logic (RWL):** The Read/Write logic accepts

inputs from the bus and generates control signals for the other functional blocks of the design. The inputs A0 and A1 selects one of the counters and the control word register. The input rd_ is the read-control signal, wr_ is the write-control signal and cs is an input enable (active low) for reading or writing.

**Control Word Register(CWR):** The CWR contents are used to determine the operation of the counters.

**Counters (CNR):** The counters consist of a status register, counter register, output latch, counting element and control logic. The control logic interprets the CWR outputs to determine the operations of the counter. The counting element is a 16-bit, presettable synchronous-down counter. The counting registers stores the written data to the latched outputs. When the counter is read, the outputs are read from the latches. The counters are read from or written to directly by selecting with A0 and A1, and asserting rd_ or wr_.

**Operation:** The counters are programmed by writing a control word to the CWR, followed by an initial count. The format of the CWR is shown below:

SC1, SC0: CNR selection input.
M2, M1, M0: determines the mode (optional mode0 for the project, mode0 is described below).
RW1, RW0: determines which significant byte to operate on.

| SC1 | SC0 | RW1 | RW0 | M2 | M1 | M0 | X |
|---|---|---|---|---|---|---|---|

**Mode0 (Interrupt on Terminal Count):** Out is initially low until the counter reaches zero. Out then goes high, until a new count or a new mode0 control word is written into the counter. Each counter is enabled by its gate signal (active high).

### Inputs

DS: Square wave signal from vehicle driveshaft
CK: Stable reference clock signal
REF: Binary number representing desired vehicle speed

### Outputs

SU: Speed up control command
SD: Slow down control command

### Operation

During the positive half-cycle of CK, the circuit counts the number of driveshaft (DS) pulses, compares this sum against the desired speed (REF), and determines whether to issue a SU or SD signal or no signal, if the speed is correct. The SU and SD signals control the vehicle during the negative half-cycle of CK.

5 usec

CK

compare speed with REF
disable speed control

disable counting logic,
issue SD, SU

**Figure 7.43** Automatic cruise control (ACC)

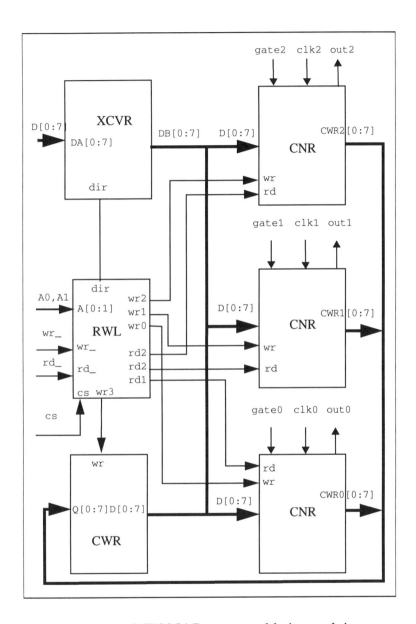

**Figure 7.44** AMD8254 Programmable interval timer

## Source Listing

### Adder Without Resource Sharing

```verilog
module adder (A,B,C,D,clk,reset,out);

    output[0:7] out;
    input[0:7] A,B,C,D;
    input clk,reset;
    reg [0:1] fsm, fsm_N;
    wire reset;
    reg[0:7] out;

    parameter state0 = 0, state1 = 1,
              state2 = 2;

    always @(posedge clk or posedge reset)
        begin:adder_reset
        if (reset) begin
            fsm = 0;
        end
        else begin
            fsm = fsm_N;
        end
    end

    always @(fsm)
        begin : adder_func
        out = 0;
        fsm_N = fsm;


        casex(fsm)
        state0: begin
            out = 0;
            fsm_N = state1;
        end
        state1: begin
            out = A + B;
            fsm_N = state2;
        end
        state2: begin
            out = C + D;
            fsm_N = state0;
        end
        default: begin
            out = 8'bx ;
            fsm_N = 'bx ;
        end
    endcase
```

```
        end
endmodule
```

## Adder Using Resource Sharing

```
module adder (A,B,C,D,clk,reset,out);

    output[0:7] out;
    input[0:7] A,B,C,D;
    input clk,reset;
    reg [0:1] fsm, fsm_N;
    wire reset;
    reg[0:7] out;

    parameter state0 = 0, state1 = 1,
              state2 = 2;

    function [0:7] outbus;
        input [0:7] IN1, IN2 ;

        begin : functionality
                outbus = IN1 + IN2;
        end
    endfunction

    always @(posedge clk or posedge reset)
        begin:adder_reset
        if (reset) begin
                fsm = 0;
        end
        else begin
                fsm = fsm_N;
        end
    end

    always @(fsm)
        begin : adder_func
        out = 0;
        fsm_N = fsm;

        casex(fsm)
        state0: begin
                out = 0;
                fsm_N = state1;
        end
        state1: begin
                out = outbus(A, B) ;
                fsm_N = state2;
        end
```

```
            state2: begin
                    out = outbus (C, D);
                    fsm_N = state0;
            end
            default: begin
                    out = 8'bx ;
                    fsm_N = 'bx ;
            end
            endcase
      end

endmodule
```

## Primitive Implementation for 4-bit to 1-bit Multiplexor

```
primitive MUX_4_2 (Y, D0, D1, D2, D3, S0, S1);
    input D0, D1, D2, D3, S0, S1;
    output Y;
    table // D0 D1 D2 D3 S0 S1 : Y
       0 ? ? ? 0 0 : 0 ;
       1 ? ? ? 0 0 : 1 ;
       ? 0 ? ? 1 0 : 0 ;
       ? 1 ? ? 1 0 : 1 ;
       ? ? 0 ? 0 1 : 0 ;
       ? ? 1 ? 0 1 : 1 ;
       ? ? ? 0 1 1 : 0 ;
       ? ? ? 1 1 1 : 1 ;
       0 0 0 0 ? ? : 0 ;
       1 1 1 1 ? ? : 1 ;
       0 0 ? ? ? 0 : 0 ;
       1 1 ? ? ? 0 : 1 ;
       ? ? 0 0 ? 1 : 0 ;
       ? ? 1 1 ? 1 : 1 ;
       0 ? 0 ? 0 ? : 0 ;
       1 ? 1 ? 0 ? : 1 ;
       ? 0 ? 0 1 ? : 0 ;
       ? 1 ? 1 1 ? : 1 ;
    endtable
endprimitive
```

## Inefficient 4-bit to 16-bit decoder

```
/* A four to 16 decoder 4-bit to 16-bit deecoder
        with enable logic*/

module d4x16 (inp, clk, reset, out,enand,
              enor, enxor);
    input [0:3] inp;
    input reset, clk;
```

```
        output [0:15] out;
        output enand, enor, enxor ;
        reg [0:15] out;
        reg enand, enor, enxor ;

        always @ (posedge clk) begin
              if (reset) begin
                    out = 0 ;
                    enand = 0 ;
                    enor = 0 ;
                    enxor = 0 ;
              end
              else begin
                    out = 0;
                    out[inp] = 1 ; // simpler modeling!
                    // or out = 16'b8000 >> inp ;
                    enand = &(out);
                    enor = !(out);
                    enxor = ^(out);
              end
        end

    endmodule
```

## Efficient 4-bit 16-bit decoder

```
    /* A four to 16 decoder with enable logic */

    module d4x16 (inp, clk, reset, out,
                    enand, enor, enxor);
         input [0:3] inp;
         input reset, clk;
         output [0:15] out;
         output enand, enor, enxor ;
         reg [0:15] out;
         wire enand, enor, enxor ;

         always @ (posedge clk) begin
               if (reset)
                    out = 16'b0 ;
               else begin
                    out = 0;
                    case (inp) // directives to tool may
                      0: out[0] = 1; // be used as comments
                      1: out[1] = 1;
                      // .....
                      // .....
                     15: out[15] = 1;
```

```
            end
    end

    assign enand = &(out);
    assign enor  = |(out);
    assign enxor = ^(out);

endmodule
```

## Timer Functional Model

```
module timer (green, yellow, red, clk, reset);
    // port declarations
    output green;
    output yellow;
    output red;
    input clk;
    input reset;

    wire [5:1] count;
    wire start;

    state state (start, green, yellow,
            red, clk, reset, count);
    counter counter (count, start, clk, reset);

endmodule
```

## Counter Model

```
module counter (count, start, clk, reset);

    input clk;
    input reset;
    input start;
    output [5:1] count;

    wire clk;
    reg [5:1] count_N;
    reg [5:1] count;

    // reset operation

    // count operation
    always @ (posedge clk or posedge reset)
        begin : counter_S
        if (reset) begin
            count = 0;
            // reset logic for the block
```

```
                    end
            else begin
                    count = count_N;
            end
    end

    always @ (count or start)
            begin : counter_C
            count_N = count;
            // initialize outputs of the block
            if (start) count_N = 1;
            else count_N = count + 1;
    end
endmodule
```

## State Model

```
module state (start, green, yellow,
              red, clk, reset, count);

    output green;
    output yellow;
    output red;
    output start;
    input clk;
    input reset;
    input [5:1] count;

    reg green;
    reg yellow;
    reg red;
    reg start;
    reg [5:1] count_N;
    reg [2:1] fsm_N;
    reg [2:1] fsm;

    // automaton fsm
            // clock registers of the block
    always @ (posedge clk or posedge reset)
            begin : fsm_S
            if (reset) begin
                    fsm = 0;
            end
            else begin
                    fsm = fsm_N;
            end
    end

    always @ (fsm or count)
            // combinational logic
```

```verilog
            begin : fsm_C
            parameter [2:1] init = 0, g = 1,
                            y = 2, r = 3;
            // initialize outputs of the block
            red = 0;
            fsm_N = fsm;
            green = 0;
            start = 0;
            yellow = 0;

            // user specified logic for fsm
            casex (fsm)
                init: begin
                    start = 1;
                    fsm_N = g;
                end
                g: begin
                    green = 1;
                    if (count == 25) begin
                        start = 1;
                        fsm_N = y;
                    end
                end
                y: begin
                    yellow = 1;
                    if (count == 2) begin
                        start = 1;
                        fsm_N = r;
                    end
                end
                r: begin
                    red = 1;
                    if (count == 15) begin
                        start = 1;
                        fsm_N = g;
                    end
                end
                default: begin
                    red = 'bx;
                    fsm_N = 'bx;
                    green = 'bx;
                    start = 'bx;
                    yellow = 'bx;
                end
            endcase
        end

endmodule
```

# AMD2910 Models

## Register Model

```verilog
'define REGDATA 2'b00
'define REGONEBIT 2'b01

module register (R, ZERO, register_cnt, D,
                 RLD_, CP);
   output [11:0] R;
   output ZERO;

   input [11:0] D;
   input [1:0] register_cnt;
   input    RLD_, CP;

   reg   [11:0] R_IN, R ;
   reg ZERO;

   always @ ( register_cnt or D or RLD_ )
        begin if ( RLD_ == 1'b0 )  // incomplete if-latch
             R_IN = D ;
          case ( register_cnt )
            'REGDATA :
                R_IN = D ;
            'REGONEBIT :
                R_IN = D  - 12'b1 ;
            default :
                R_IN = 12'bx ;
          endcase
   end

   always @ (posedge CP) begin
        R = R_IN ;
   end

   always @ ( R ) begin
        ZERO = ~( |( R ));
   end
endmodule
```

## Stack Model

```verilog
'define HOLD 2b'00
'define POP 2b'01
'define PUSH 2b'10
'define CLEAR 2b'11
module stack (F, FULL_, stack_cnt, UPC, CP);
output [11:0] F;
output FULL_;
```

```verilog
        input [1:0] stack_cnt;
        input [11:0]UPC;
        input CP;

        reg [11:0] F;
        reg [8:0] stack_BIT0, stack_BIT1,
                  stack_BIT2, stack_BIT3,
                  stack_BIT4, stack_BIT5,
                  stack_BIT6, stack_BIT7,
                  stack_BIT8, stack_BIT9,
                  stack_BIT10, stack_BIT11;
        reg FULL_;
        reg [3:0] ADDR;

        always @ (posedge CP)
         case (stack_cnt)
                // may use special information such as
                // parallel case
                'HOLD :
                 ;// HOLD
                'POP :
                 if (ADDR != 0) ADDR = ADDR - 1;// POP
                'PUSH : begin// PUSH
                 ADDR = ADDR + 1;
                 stack_BIT0[ADDR] = UPC[0];
                 stack_BIT1[ADDR] = UPC[1];
                 stack_BIT2[ADDR] = UPC[2];
                 stack_BIT3[ADDR] = UPC[3];
                 stack_BIT4[ADDR] = UPC[4];
                 stack_BIT5[ADDR] = UPC[5];
                 stack_BIT6[ADDR] = UPC[6];
                 stack_BIT7[ADDR] = UPC[7];
                 stack_BIT8[ADDR] = UPC[8];
                 stack_BIT9[ADDR] = UPC[9];
                 stack_BIT10[ADDR] = UPC[10];
                 stack_BIT11[ADDR] = UPC[11];
                end
                'CLEAR :
                 ADDR = 0 ; // CLEAR
         endcase

    if ( ADDR >= 9 ) begin
            FULL_ = 1'b0;
            ADDR = 9 ;
            end
    else
            FULL_ = 1'b1;
            F[0] = stack_BIT0[ADDR];
            F[1] = stack_BIT1[ADDR];
            F[2] = stack_BIT2[ADDR];
```

```
            F[3]  = stack_BIT3[ADDR];
            F[4]  = stack_BIT4[ADDR];
            F[5]  = stack_BIT5[ADDR];
            F[6]  = stack_BIT6[ADDR];
            F[7]  = stack_BIT7[ADDR];
            F[8]  = stack_BIT8[ADDR];
            F[9]  = stack_BIT9[ADDR];
            F[10] = stack_BIT10[ADDR];
            F[11] = stack_BIT11[ADDR];
      end

endmodule
```

## Incrementor Model

```
'define UPCINC 1'b0
'define UPCDEC 1'b1

module incrementor (UPC, incrementor_cnt,
              Y_IN, CI, CP);
   output [11:0] UPC;
   input  [11:0] Y_IN;
   input  incrementor_cnt;
   input   CI, CP;

   reg    [11:0] UPC;
   reg    [11:0] UPC_IN;

   always @ (Y_IN or incrementor_cnt or CI)
         begin
           case ( incrementor_cnt )
                'UPCINC :
                      UPC_IN = Y_IN + CI;
                'UPCDEC :
                      UPC_IN = 0;
                default :
                      UPC_IN = 0 ;
           endcase
      end

      always @ ( posedge CP ) begin
            UPC = UPC_IN ;
      end

endmodule
```

## 11-bit Multiplexor

```
'define DATA 2'b00
```

```verilog
'define REG 2'b01
'define STACKIN 2'b10
'define UPCOUT 2'b11
module mux (Y_IN, mux_cnt, D, R, F, UPC);

    output [11:0] Y_IN;
    input  [1:0]  mux_cnt;
    input  [11:0] D, F, R, UPC;

    reg    [11:0] Y_IN;

    always @ ( mux_cnt or D or R or F or
                UPC ) begin
     case ( mux_cnt )
         'DATA :
                Y_IN = D ;
         'REG :
                Y_IN = R ;
         'STACKIN :
                Y_IN = F ;
         'UPCOUT :
                Y_IN = UPC;
     endcase
    end

endmodule
```

## PLA Model

```verilog
`define JZ 4'b0000
`define CJS 4'b0001
`define JMAP 4'b0010
`define CJP 4'b0011
`define PUSHIN 4'b0100
`define JSRP 4'b0101
`define CJV 4'b0110
`define JRP 4'b0111
`define RFCT 4'b1000
`define RPCT 4'b1001
`define CRTN 4'b1010
`define CJPP 4'b1011
`define LDCT 4'b1100
`define LOOP 4'b1101
`define CONT 4'b1110
`define TWB 4'b1111

`define INCNT 1'b0
`define INCHI 1'b1

`define STKCNT 2'b00
```

```verilog
`define DATCNT 2'b01
`define REGCNT 2'b10
`define MUXCNT 2'b11

module pla (PL_, MAP_, VECT_, register_cnt,
            mux_cnt, stack_cnt, incrementor_cnt,
                ZERO, I, CC_, CCEN_);
    output PL_, MAP_, VECT_;
    output [1:0] register_cnt, mux_cnt,
                stack_cnt;
    output incrementor_cnt;
    input  [3:0] I;
    input  ZERO, CC_, CCEN_;

    reg   [1:0]  register_cnt, mux_cnt,
                stack_cnt;
    reg    incrementor_cnt;
    reg    PASS;
    reg    PL_;
    reg    MAP_;
    reg    VECT_;

    always @ ( CC_ or CCEN_ ) begin
            PASS = ~CC_ | CCEN_ ;
    end

    always @ ( I or ZERO or PASS ) begin
        stack_cnt = 'STKCNT;// NC
        mux_cnt = 'MUXCNT;// UPC
        register_cnt = 'REGCNT;// NC
        incrementor_cnt = 'INCNT ;
        PL_ = 'INCNT ;// LOW
        MAP_ = 'INCHI ;
        VECT_ = 'INCHI ;

        case ( I )
        'JZ : begin
         stack_cnt = 'MUXCNT;// CLEAR
         incrementor_cnt = 'INCHI ;// CLEAR
        end
        'CJS : begin
          if ( PASS ) begin
                stack_cnt = 'REGCNT;// PUSH
                mux_cnt = 'STKCNT;// D
          end
        end
        'JMAP :begin
         mux_cnt = 'STKCNT;// D
         MAP_ = 'INCNT ;// LOW
        end
```

```
'CJP :begin
  if ( PASS ) begin
       mux_cnt = 'STKCNT;// D
  end
end
'PUSHIN :begin
 stack_cnt = 'REGCNT;// PUSH
 if ( PASS ) begin
       register_cnt = 'STKCNT;// LOAD
 end
end
'JSRP :begin
 stack_cnt = 'REGCNT;// PUSH
 if ( PASS ) begin
       mux_cnt = 'STKCNT;// D
 end
 else
       mux_cnt = 'DATCNT;// R
end
'CJV :begin
 if ( PASS ) begin
       mux_cnt = 'STKCNT;// D
 end
       VECT_ = 'INCNT ;
end
'JRP :begin
 if ( PASS ) begin
       mux_cnt = 'STKCNT;// D
 end
 else
       mux_cnt = 'DATCNT;// R
end
'RFCT :begin
 if ( ZERO ) begin
       mux_cnt = 'REGCNT;// F
       register_cnt = 'DATCNT;// DEC
 end
 else
      stack_cnt = 'DATCNT;// POP
end
'RPCT :begin
 if ( ZERO ) begin
       mux_cnt = 'STKCNT;// D
       register_cnt = 'DATCNT;// DEC
 end
end
'CRTN :begin
 if ( PASS ) begin
       mux_cnt = 'REGCNT;// F
       stack_cnt = 'DATCNT;// POP
```

```verilog
                end
            end
            'CJPP :begin
             if ( PASS ) begin
                    mux_cnt = 'STKCNT;// D
                    stack_cnt = 'DATCNT;// POP
             end
            end
            'LDCT :begin
                    register_cnt = 'STKCNT;// LOAD
            end
            'LOOP :begin
             if ( PASS )
                    stack_cnt = 'DATCNT;// POP
             else
                    mux_cnt = 'REGCNT;// F
            end
            'CONT :begin
            end
            'TWB : begin
             if ( PASS & ZERO ) begin
                    stack_cnt = 'DATCNT;// POP
             end
             else if ( PASS & ~ZERO ) begin
                    stack_cnt = 'DATCNT;// POP
                    register_cnt = 'DATCNT;// DEC
             end
             else if ( ~PASS & ZERO ) begin
                    stack_cnt = 'DATCNT;// POP
                    mux_cnt = 'STKCNT;// D
             end
             else if ( ~PASS & ~ZERO ) begin
                    mux_cnt = 'REGCNT;// F
                    register_cnt = 'DATCNT;// DEC
             end
            end
          endcase
    end

  endmodule
```

## Tristate Models

```verilog
  module out (Y, Y_IN, OE_);
      output [11:0] Y;
      input  [11:0] Y_IN;
      input     OE_;

/* the "Tristate" module represents a typi
     cal ASIC vendor library tri-state output
```

```
         cell, with the parameters (tristated_
         data_output, data_input, tristate_en
                    able_input) */


         TRISTATE TRI11 (Y_IN[11], OE_, Y[11]),
                  TRI10 (Y_IN[10], OE_, Y[10]),
                  TRI9 (Y_IN[9], OE_, Y[9]),
                  TRI8 (Y_IN[8], OE_, Y[8]),
                  TRI7 (Y_IN[7], OE_, Y[7]),
                  TRI6 (Y_IN[6], OE_, Y[6]),
                  TRI5 (Y_IN[5], OE_, Y[5]),
                  TRI4 (Y_IN[4], OE_, Y[4]),
                  TRI3 (Y_IN[3], OE_, Y[3]),
                  TRI2 (Y_IN[2], OE_, Y[2]),
                  TRI1 (Y_IN[1], OE_, Y[1]),
                  TRI0 (Y_IN[0], OE_, Y[0]);

endmodule
```

## High Impedance Tristate Model

```
module TRISTATE (Y_IN, OE_, Y);
   output  Y;
   input   Y_IN, OE_;
   reg     Y;

   always @ (OE_) begin
    if (OE_)
        Y = Y_IN;
    else
        Y = 1'bz; // assign high-impedance
   end

endmodule
```

## AMD2910 Structural Model

```
module AMD2910 (VECT_, MAP_, PL_, FULL_, Y,
         CP, OE_, RLD_, CI,CCEN_, CC_, I, D);

   output  VECT_, MAP_, PL_, FULL_;
   output  [11:0] Y;
   input    CP, OE_, RLD_, CI, CCEN_, CC_;
   input   [3:0] I;
   input   [11:0] D;

   wire    ZERO, LOAD, DECR, HOLD, CLEAR,
              POP, PUSH;
   wire    [11:0] R, Y_IN, F, UPC;
```

```verilog
        wire    [1:0] register_cnt, stack_cnt,
                mux_cnt;
        wire    incrementor_cnt;

        register register (R, ZERO, register_cnt,
                D, RLD_, CP);
        stack  stack (F, FULL_, stack_cnt, UPC, CP);
        incrementor incrementor (UPC,
                incrementor_cnt,Y_IN, CI, CP);
        mux    mux (Y_IN, mux_cnt, D, R, F, UPC);
        pla    pla (PL_, MAP_, VECT_, register_cnt,
               mux_cnt,stack_cnt,incrementor_cnt,
               ZERO, I, CC_, CCEN_);
        out    out (Y , Y_IN, OE_);

endmodule
```

CHAPTER

# 8

# Modeling a Floppy Disk Subsystem

In this chapter we provide a complete example of a floppy disk subsystem (FDS) model in Verilog. Such a model may be needed when you perform a full system simulation and want to simulate the execution of code which accesses the disk. The FDS model would typically be used to perform a full functional simulation during the development of a CPU board. This example demonstrates the modeling of asynchronous systems, I/O buses, timing constraints and other techniques of writing large models.

**Functional Description**

Figure 8.1 shows the configuration of a typical system that includes a processor and a disk. The CPU communicates with a floppy disk controller (FDC) chip (e.g., the WD57C65) through the data and control buses. The FDC, in turn, communicates with the floppy disk drive (FDD) through two serial data lines and various control and status lines. The CPU sends commands to the FDC by writing a sequence of bytes to one of its internal registers. When the FDC receives a command, it begins to execute the command. Some commands involve reading data bytes from the CPU and sending them serially to the disk drive or reading data from the disk drive on the serial line and sending them as parallel bytes

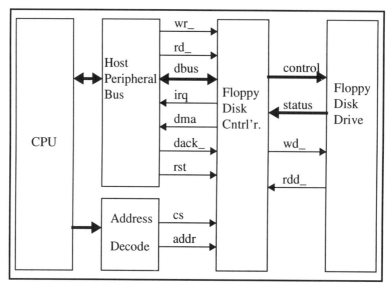

**Figure 8.1** A typical CPU and floppy disk subsystem

to the CPU. The input/output pins of the controller are shown in Figure 8.2.

Since our purpose is not to design a floppy disk controller but only to develop a model in order to test the CPU design, the FDC model is greatly simplified. A real FDC chip might be able to control multiple

```
wr_         - write signal (I)
rd_         - read signal (I)
dbus        - data bus (I/O)
dma         - dma request (O)
irq         - interrupt request (O)
dack_       - dma acknowledge (I)
rst         - reset (I)
cs_         - chip select (I)
addr        - address (I)

moen_       - motor enable (O)
tr00_       - track 0 (I)
idx_        - index hole indicator (I)
rdd_        - serial data in (I)
wd_         - serial data out (O)
dirc_       - direction of head movement (O)
```

**Figure 8.2** Pinout description of the controller

floppy disk drives, each drive with two heads. In this example, however, the FDC controls a single one-sided drive. Most of the signals on the CPU side are similar to those of a real FDC chip, but the signals between the FDC and the FDD have been significantly modified.

In a real controller/drive system, data is transferred on two serial lines, one for input and one for output. Because of the conflicting requirements of high data density and high reliability, the encoding techniques used to serialize and deserialize the data can be quite complex. Moreover, since the spinning speed of the drive cannot be controlled very accurately, clock information must be embedded in the data, which complicates the encoding methods even further. Recovering the clock from the data requires designing a data separator and phase-locked loop (PLL) circuit, not a trivial task.

The simplified model uses parallel 8-bit buses to transfer information between the controller and the drive and a special channel to carry the clock. This not only simplifies the coding of the model, but also speeds up the simulation while maintaining a faithful disk subsystem model with respect to the host.

## Operation of the Disk Subsystem

The FDS is depicted in Figure 8.3. It has three sub modules: a timing checker, an FDC, and an FDD.

The FDC receives commands from the CPU and executes the commands. To reduce the complexity of the model, our FDC can accept only six commands, which have been greatly simplified. The FDC commands are shown in Figure 8.4.

The READ_DATA and WRITE_DATA commands read and write a single sector from the current track. The sectors on each track are ordered sequentially from 0 to SECTORS_PER_TRACK-1. FORMAT_TRACK marks the current track as formatted and writes a filler byte for all the sectors in the track. To execute a SEEK command, the FDC sets the dirc_ output and sends step pulses to the disk drive to move the disk head to the desired track. The RECALIBRATE command retracts the disk head to track 0. This is done by setting the dirc_ output and sending step pulses until the outermost track (tr00_) is active, indicating that the head is fully retracted.

# Digital Design and Synthesis with Verilog HDL

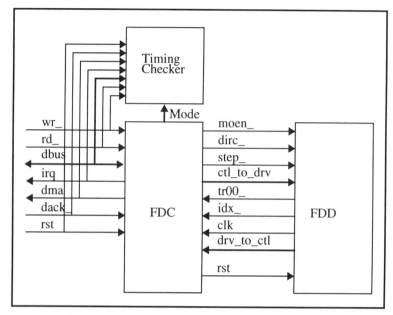

**Figure 8.3** Floppy disk subsystem block diagram

```
Cmd #    Command name    #of bytes        Cmd code
-----------------------------------------------------
1        READ_DATA       2                'b0110
2        WRITE_DATA      2                'b0101
3        FORMAT_TRACK    2                'b1101
4        SEEK            2                'b1111
5        RECALIBRATE     1                'b0111
```

**Figure 8.4** Floppy disk controller commands

The following sections describe the components of the FDS model: the timing checker, the FDC and the FDD.

## The Timing Checker

The timing checker is in effect not a functional part of the FDS, but it is used in modeling to detect illegal input patterns. Since these inputs are presumably generated by the CPU model, any such illegal input indicates a possible error in the design of the CPU. As it turns out, the timing checker is also useful for debugging the FDS.

The timing checker code contains two types of loops. The first type records only the time at which a signal changes its state, and the second checks for minimum or maximum delay constraints such as setup, hold, and data width times. A sample of the two types of loops is shown in Figure 8.5. The first loop in the example records the times at which the rd_ signal changes from low to high or from high to low, and the second loop checks for minimum read data width.

```
time rd_high;
wire programmed_io = (idle_mode || command_mode) && !cs_;
always @rd_ begin
        if (rd_ == 1)
                rd_high = $time;
        else if (rd_ == 0)
                rd_low = $time;
        else if (programmed_io) illegal_signal ("rd_", rd_);
end
always @(posedge rd_) if (programmed_io)
  checkmintiming ("tRR", tRRmin, rd_low);
```

**Figure 8.5** Recording and checking timing violations

The important work is done in the `checkmintiming` and `checkmaxtiming` tasks. The `checkmintiming` task (Figure 8.6) verifies that the time difference between two events is greater than some minimum value and the `checkmaxtiming` task is its counterpart for checking maximum delays. A third task, the `illegal_signal` task, just prints an error message and stops the simulation if an illegal signal is encountered.

## The Floppy Disk Controller

The controller is divided logically into two sections: the host interface and the drive interface. The host interface section operates between the FDC and the host CPU. It accepts commands from the host and initiates their execution, sends and receives data bytes from the CPU in EXECUTE mode, and sends status information to the CPU in IDLE mode. The drive interface transfers data between the controller and the disk drive during execution, sends control signals to the drive, and receives status information from the drive.

```
task checkmintiming;
input message;
input tdiff;
input prevtime;
...
begin
 if ($time - prevtime < tdiff) begin
   $display("%m:Timing violation: %s, %0d-%0d<%0d",
       message, $time, prevtime, tdiff);
   'STOP;
 end
end
endtask
```

**Figure 8.6** The checkmintiming task

The FDC has two internal registers: `main_stat_reg` and `main_data_reg`. The `main_data_reg` is a general purpose 8-bit register, which can be read and written by the processor. In COMMAND mode, the command bytes must be written one-by-one into `main_data_reg`. The `main_data_reg` can also be read in COMMAND mode. The `main_stat_reg` is a read-only register which contains the status of the FDC. The meaning of the various bits in the status register is given in Figure 8.7. At the end of each command, the

```
parameter
 ST_COMPLETE  = 0,  // The command was completed
 ST_NOINDEX   = 1,  // Could not find the index mark
 ST_ILLEGAL   = 2,  // Illegal command
 ST_CHECKSUM  = 3,  // Bad checksum
 ST_OVERRUN   = 4,  // Overrun while sending to the host
 ST_UNDERRUN  = 5;  // Underrun while reading from the host
reg[7:0] main_stat_reg;
```

**Figure 8.7** The status bits

CPU should read the status register and confirm the successful completion of the command.

**Programmed I/O and DMA transactions**

The controller can communicate with the CPU using two different protocols: programmed I/O and direct memory access (DMA). The main difference between the two is that programmed I/O requests are initiated by the CPU, which uses the cs_ line to select the FDC device. Typically cs_ is decoded from a memory address or address range. DMA transfers are initiated by the FDC, and use the dma and dack_ signals for

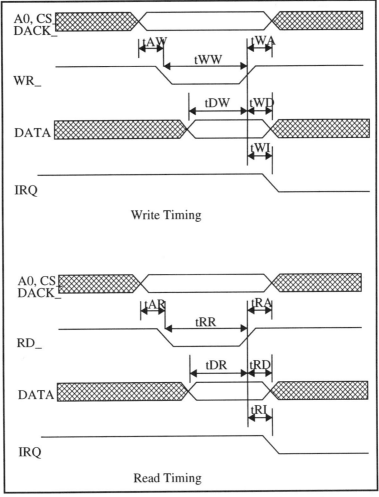

**Figure 8.8** Waveforms for a programmed I/O operation

# Digital Design and Synthesis with Verilog HDL

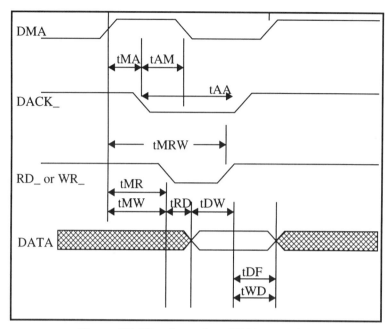

**Figure 8.9** Waveforms for a DMA operation

handshaking. Another difference is that in programmed I/O, the addr input selects which internal register of the FDC is read or written. In DMA mode, addr is not used.

The waveforms for programmed I/O and DMA transactions are shown in Figure 8.8, and Figure 8.9 respectively.

The CPU starts a programmed I/O transaction by asserting cs_ and deasserting dack_ and setting addr. The addr input selects the desired internal register to be read or written. For a programmed read operation, the CPU asserts rd_ and reads the data off the dbus. Finally, the CPU deasserts rd_ and then deasserts cs_. The write operation is similar, except that the wr_ signal is asserted instead of the rd_, and the dbus is written by the CPU rather than read.

A DMA transaction starts when the FDC asserts the dma signal. When the CPU responds by asserting dack_, the FDC clears dma and waits for a rd_ or wr_ to be asserted. If rd_ is asserted then followed by a read operation, (i.e., the CPU reads from the FDC) and if wr_ is

```
always @(negedge rd_) begin
    if (dack_ == 1 && cs_ == 0) begin // Programmed read
        if (addr == MAIN_STAT)
            dbus_reg = main_stat_reg;
        else
            dbus_reg = main_data_reg;
        ->do_read_byte;
    end
    else if (~dack_) begin  // DMA read
        dbus_reg = byte_tofrom_host;
        #tDF dbus_reg = 8'hZZ;
    end
end
```

**Figure 8.10** Programmed I/O and DMA read transactions

```
always @(posedge wr_) begin
    if (dack_ == 1 && cs_ == 0) begin // Programmed write
        if (idle_mode) begin
            current_command = 0;
            mode = 'COMMAND;
        end
        main_data_reg = dbus;
        current_command = current_command + 1;
        command_array[current_command] = dbus;
        if (current_command == 1) new_command = dbus;
            if (current_command == cmd_bytes[new_command])
                begin
                mode = 'EXECUTE;
                main_stat_reg = 0;
                execute;
                main_stat_reg[ST_COMPLETE] = 1;
                mode = 'IDLE;
                irq = 1;
                ->unload_head;
            end
    end
end

always @(posedge wr_) begin
    if (~dack_)      // DMA write
        byte_tofrom_host = dbus;
end
```

**Figure 8.11** Programmed I/O and DMA write transactions

asserted, then a write operation is performed. The transaction ends when the `rd_` or `wr_` signal is deasserted, `dack_` is deasserted, and the `dbus` is tristated. The code that initiates programmed and DMA read transactions is shown in Figure 8.10, and the code that initiates write transactions is shown in Figure 8.11.

In IDLE mode and in COMMAND mode, the controller accepts programmed I/O requests from the CPU. In EXECUTION mode the controller initiates one DMA request for each byte to be transferred between the host CPU and the drive.

**Processing the Controller Commands**

The controller is always in one of three modes: IDLE, COMMAND, or EXECUTE, as shown in the code segment of Figure 8.12

After a reset or after completing the execution of a command, the FDC is in IDLE mode. In this mode the FDC waits for the host to send command bytes using programmed I/O. The first byte of a command defines the command type and implies the command length. Commands can be one or two bytes long. When the FDC receives the first byte of a new command, it switches to COMMAND mode and remains in that state until it receives the last byte of the command. When the last byte of the command is received, the FDC switches to EXECUTE mode and starts executing the command. When it finishes executing the command,

```
'define IDLE    2'b00
'define COMMAND 2'b01
'define EXECUTE 2'b10

reg [1:0] mode;
wire idle_mode       = (mode == 'IDLE);
wire command_mode    = (mode == 'COMMAND);
wire execute_mode    = (mode == 'EXECUTE);

always @(posedge rst) begin
        ...
        mode = 'IDLE;
        ...
end
```

**Figure 8.12** The three controller modes

the FDC switches back to IDLE mode and stays in this mode until a new command byte is transmitted by the host CPU.

The execution of the command might entail controlling the disk drive, setting parameters in the controller, or transferring data between the disk and the CPU. When the FDC completes the execution of a command, it initiates an interrupt request by setting the irq signal, and the status of the FDC is reset to IDLE mode. The CPU can reset the interrupt request by reading the status register. A successful completion of the command is indicated by status = 8'h01.

```
task execute; begin
        case (new_command)
        READ_DATA,
        WRITE_DATA,
        FORMAT_TRACK: begin
                read_write (new_command, command_array[2]);
        end
        SEEK: begin : seek_block
                integer diff, i;
                if (current_track > command_array[2]) begin
                    diff = current_track - command_array[2];
                        dirc_ = 1;
                end
                else begin
                    diff = command_array[2] - current_track;
                        dirc_ = 0;
                end
                for (i = 0; i < diff; i = i + 1)
                        one_step;
                current_track = command_array[2];
        end
        RECALIBRATE: begin
                dirc_ = 1;
                while (tr00_)
                        one_step;
                current_track = 0;
        end
        default: begin
                main_stat_reg[ST_ILLEGAL] = 1;
                $display ("Illegal command");
                'STOP;
        end
        endcase
end
endtask
```

FDC execute

# Digital Design and Synthesis with Verilog HDL

When the controller receives a full command, it invokes the `execute` task (). This task is implemented as a big case statement. It directly executes the commands SEEK and RECALIBRATE, but delegates the execution of the commands READ_DATA, WRITE_DATA, and FORMAT_TRACK to another task: read_write. The SEEK command generates step pulses until the head has moved to the desired track, and the RECALIBRATE command generates step pulses until the head is fully retracted and is positioned at track 0.

```
task read_write;
...
begin
   ->load_head;
   wait (head_is_loaded);
   main_stat_reg[ST_NOINDEX] = 1;
   @(negedge idx_)
   main_stat_reg[ST_NOINDEX] = 0;
   $display ("Detected idx_ (command %0s) at time %0t",
      cmd_names [operation], $time);
   case (operation)
   FORMAT_TRACK: begin
      for (sector = 0; sector < SECTORS_PER_TRACK;
         sector = sector + 1) begin
         ...
         // Format one sector
         ...
      end
   end
   WRITE_DATA, READ_DATA: begin
   begin : search_sector
      for (sector = 0; sector < SECTORS_PER_TRACK;
         sector = sector + 1) begin
         ...
         // If found then disable search_sector
         // Else skip this sector
         ...
      end
      $display ("Could not find the sector");
      'STOP;
   end     // search_sector block
   if (operation == WRITE_DATA)
      write_data;
   else
      read_data;
   end
   endcase
end
endtask
```

**Figure 8.14** The read_write task

The read_write task is shown in Figure 8.14. It starts by loading the head and waiting for the synchronizing signal idx_. To perform the FORMAT_TRACK command, the FDC model sends bytes to the FDD, one sector at a time, until all the sectors of the track have been written. Each sector has a header (the ordinal sector number in the track), a data section (filled by some filler bytes), and a tail (a checksum of all the data bytes and the sector head). To perform a READ_DATA or WRITE_DATA command, the FDC model searches for the appropriate sector, and then calls yet another task: write_data for writing, or read_data for reading.

The read_data and write_data tasks initiate the actual DMA transfer of bytes between the FDD and the host CPU. Each transfer has to be synchronized with the clock signal from the FDD. In order to avoid race conditions, all data is sent on the negative edge of the clock and is sampled on the positive edge of the clock. If the FDC detects a data overrun during a WRITE_DATA or READ_DATA operation, it sets the corresponding bit in the status register.

```
moen_     -- Motor enable (I). When active, then after some
             delay the disk starts rotating and sends idx_
             pulses, one per rotation
dirc_     -- Direction (I). When active, then each step
             pulse moves the head one track forwards,
             otherwise, each step pulse moves the head one
             track backwards.
step_     -- Step (I). Each pulse moves the head one cylinder
             in a direction specified by dirc_.
byte_in   -- Byte_in (I). Input byte from the controller.
tr00_     -- Track zero (O). When active, it indicates that
             the head is retracted to track zero.
idx_      -- Index (O). Every revolution of the diskette,
             the drive sends a pulse to the controller. This
             indicates the beginning of the track.
clk       -- Clock (O). Used to synchronize the drive with
             thecontroller. In a real floppy drive the clock
             is derived from the data.
byte_out  -- Byte_ out (O). Output byte to the controller.
rst       -- Reset (I).
```

**Figure 8.15** FDD I/O signals

## The Floppy Disk Drive

The disk drive model simulates the physical drive unit, including the actual floppy disk and the information that it contains. Figure 8.15 shows the signals which communicate between the drive and the controller.

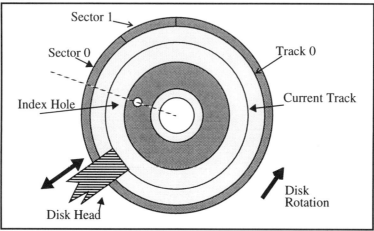

**Figure 8.16** Floppy disk

The floppy disk itself is divided into concentric cylinders. In a multisurface disk drive, each cylinder defines concentric circles, called tracks, one track per surface. Most floppy disk drives are double sided, i.e. they have two surfaces and two heads. Our model has only one surface, and therefor each cylinder corresponds to only one track. Tracks are further divided into sectors. Figure 8.16 depicts a floppy disk with its tracks and sectors.

The model parameterizes the size and speed of the disk (Figure 7.17). During debugging the various parameters can be modified in order to reduce the amount of data that has to be generated and checked. For example, you can create a disk with five tracks, three sectors per track, and five bytes per sector.

The disk head moves radially, and at each point in time the head is positioned above one track. This "current track" is the one that is read or written. When the head is positioned above track 0, the outermost track, the output `tr00_` is active. The FDD monitors the `step_` input,

# Modeling a Floppy Disk Subsystem

```
parameter
   USECS = 10,      // Time units per microsecond
   MSECS = 10000,   // Time units per millisecond
   SECTORS_PER_TRACK = 16,
   MAX_TRACK = 96,
   DATA_PER_SECTOR = 256,
   BYTES_PER_SECTOR = DATA_PER_SECTOR + 2,
   // i.e. Data bytes + sector number + checksum
   BYTES_PER_TRACK = BYTES_PER_SECTOR * SECTORS_PER_TRACK,
   BYTES_PER_DISK = BYTES_PER_TRACK * MAX_TRACK,
   TOTAL_SECTORS = SECTORS_PER_TRACK * MAX_TRACK,
   HALF_CYCLE = 2 * USECS,
   FULL_ROTATION = BYTES_PER_TRACK * 2 * HALF_CYCLE + 200;
```

**Figure 7.17** Floppy Disk Parameters

and each step_ pulse moves the head by one track in a direction determined by the `dirc_` input. Both the FDD and the FDC models maintain their own variable representing the current track. The FDC cannot read the current track of the FDD, and the only way to

```
integer current_track;
wire tr00_ = (current_track != 0);

always @(posedge step_) begin
        if (dirc_ == 1 && current_track > 0)
              current_track = current_track - 1;
        if (dirc_ == 0 && current_track < MAX_TRACK - 1)
              current_track = current_track + 1;
end
```

**Figure 8.18** Simulating disk head movement

synchronize the two variables is for the CPU to perform the RECALIBRATE command, which retracts the head to track 0.

Figure 8.18 shows the model segment that implements head movement. Each negative pulse on the `step_` input increments or decrements `current_track`, depending on the value of the direction input `dirc_`.

The floppy disk has an index hole. When the disk rotates, an LED generates a pulse every time the hole passes under it. This pulse is the `idx_` output from the FDD; it is used for synchronizing the FDC and the FDD. Each track is divided into SECTORS_PER_TRACK sectors which

```
reg rotating; // Indicates whether the disk is rotating
always begin
        #FULL_ROTATION
        ->do_idx;
        if (rotating) begin
                idx_ = 0;
                #1 idx_ = 1;
        end
end

always @moen_ begin
        rotating = 0;
        idx_ = 1;
        if (~moen_) begin
                #RAMPTIME
                if (~moen_) rotating = 1;
        end
end
```

**Figure 8.19** Simulating disk rotation

are numbered sequentially. When the FDD needs to access a sector, for example sector N, it first waits for the index signal, and then skips N-1 sectors. Figure 8.19 shows how to generate the index signal of the disk.

When the motor is on, the disk rotates and generates an index signal once per rotation. When the disk motor shuts off (moen_ == 1) and then turns back on, there is a RAMPTIME period before the signal rotating becomes active and enables the generation of index pulses.

The FDD sends data bytes to the FDC on the byte_out output and receives FDC data on the byte_in input. The transmission of bytes is synchronized by the clk signal. The clk signal is generated by the FDC and is also output to the FDD. In order to avoid race conditions, both the FDC and the FDD send the data on the negative edge of the clock and sample the data on the positive edge of the clock.

Each sector in a track is divided into a header, a body, and a tail. The header is one byte, giving the ordinal number of the sector in the track starting from 0 up to SECTORS_PER_TRACK-1. The body of the sector has DATA_PER_SECTOR data bytes, and the tail of the sector is a one-byte checksum which adds together all the bytes in the header and the body, ignoring overflow.

```
integer byte_index;
reg[7:0] checksum;
// one bit per track indicating
// whether the track is formatted
reg [MAX_TRACK-1:0] formatted;
reg [7:0] diskmem [0:BYTES_PER_DISK]; // Holds disk data
```

**Figure 8.20** Data registers for representing the disk data

If the current sector is formatted then, while the disk is rotating, the FDD continuously sends its sector data to the FDC. The FDD also continuously monitors the byte_in input. During a read operation or in IDLE mode, the FDC sets the byte_in input to undefined. If this byte is not undefined (8'hxx), the byte is written on the disk as part of the header, body, or tail of the sector. Also, if any sector header is written on the track, the track is marked as formatted.

The FDD maintains several variables and arrays to represent the data on the floppy disk. These are shown in Figure 8.20.

The vector formatted has one bit per track. When this bit is on, the corresponding sector is formatted; otherwise it is unformatted. The diskmem array represents all the data on the disk, including sector heads and tail. The data is organized sequentially, one track after the other, as are the sectors within a track. The integer byte_index points to the current byte to be written and possibly read. The value of byte_index for the beginning of the current track is calculated as

```
byte_index = current_track * BYTES_PER_TRACK;
```

The byte_index variable is incremented on every negative edge of the clock each time a byte is sent to the FDC. The checksum is a global variable which keeps the checksum for the current sector. When a checksum is read from the FDC, it is compared to the internally maintained checksum, and an error is generated if they do not match.

The loop that processes the bytes of the current track is shown in Figure 8.21. First, the clock is synchronized with the index signal by disabling the clock_gen block and byte_index is initialized. Then each sector is processed: first the sector head, then its data, and finally its tail.

```
always @do_idx begin : generate_byte
        integer sector;
        disable clock_gen;
        if (~rotating) begin
                byte_out = 8'hxx;
                disable generate_byte;
        end
        byte_index = current_track * BYTES_PER_TRACK;
        for (sector = 0; sector < SECTORS_PER_TRACK;
                                sector = sector + 1) begin
                sector_head (sector);
                sector_data;
                sector_tail;
        end
end
```

**Figure 8.21**  Processing a disk track

```
task sector_data;
integer i;
begin
        for (i = 0; i < DATA_PER_SECTOR; i = i + 1) begin
                @(negedge clk) sendbyte;
                @(posedge clk) if (byte_in !== 8'hxx)
                                getbyte;
                checksum = checksum + diskmem[byte_index-1];
        end
end
endtask

task sendbyte;
begin
        byte_out = diskmem[byte_index];
        byte_index = byte_index + 1;
end
endtask

task getbyte;
begin
        diskmem[byte_index - 1] = byte_in;
end
endtask
```

**Figure 8.22**  Processing of a sector data

Figure 8.23 shows a task that displays the disk information for debugging purposes. The sd macro is a short notation for invoking this task. This macro can be called from any Verilog naming scope because it has the full hierarchical name of the task. The macro name is kept short intentionally in order to minimize keystrokes in interactive mode.

# Modeling a Floppy Disk Subsystem

```
'define sd test_fdc.f.di.showdisk; #0 $stop;

task showdisk;
integer i, j, k, index;
begin
   for (i = 0; i < MAX_TRACK; i = i + 1)
      begin : track_block
      if (~formatted[i]) disable track_block;
      index = i * BYTES_PER_TRACK;
      $display ("Track = %0d, index = %0d", i, index);
      for (j = 0; j < SECTORS_PER_TRACK; j = j + 1) begin
         $write ("Sector %0d: ", j);
         for (k = 0; k < BYTES_PER_SECTOR; k = k + 1) begin
            $write (" %h", diskmem[index]);
            index = index + 1;
         end
         $display (";");
      end
   end
end
endtask
```

**Figure 8.23** Displaying the disk information

The processing of the sector head is very similar to the processing of sector data and the sector tail. The `sector_data` task with the two supporting tasks sendbyte and getbyte are depicted in Figure 8.22. For each byte in the sector, the FDD first sends the byte to the FDC, on the negative edge of clk. On the positive edge of `clk`, it samples the data byte from the FDC. If the sampled byte is not undefined (x), then the FDD assumes that the FDC is writing, and the FDD updates the sector data. After each byte the FDD updates the checksum for the sector.

## Testing the Subsystem

This section describes the top level test module for the FDC. The full model of `test_fdc` is shown with the rest of the modules at the end of this chapter. Here we shall just summarize the main points of the model. Figure 8.24 shows the skeleton of the top module.

The global variable receiving is set to 1 during a READ_DATA command and to 0 during a WRITE_DATA command. The block that processes DMA requests checks this variable and generates an input or an output as required. The buff array has the data for one full sector of the disk.

```
module test_fdc;
...
// If 1 then DMA transactions read from the bus
reg receiving;
// Hold one sector worth of data
reg[7:0] buff[0:BUFFSIZE-1];
...
// Instantiate three sub modules:
// timing checker,controller and drive checktiming ct(...);
disk_controller fdc (...);
disk_drive fdd (...);

// Process DMA request by the controller
always @(posedge dma) begin
        if (receiving) begin
                // Read the data byte from the bus into buff
                ...
        end
        else begin
                // Write the data byte from buff to the bus
                ...
        end
end

// Define utility tasks
...
// Main test body
initial
        // Send command #1
        // Wait for completion
        // Send command #2
        // Wait for completion
        ...
        // Display the contents of the disk
end
endmodule
```

**Figure 8.24** Test module modeling

The test_fdc module defines several utility tasks. The tasks initbuff and initbufwrite initialize buff for reading and writing respectively. The showbuf task and the sb macro are used to dump the contents of the buffer interactively. The dorst task generates a reset pulse for the controller, and the sndbyte and rcvbyte tasks send and receive one byte in programmed I/O mode. The sndcmd1 and sndcmd2 tasks send commands of length one and two respectively to the controller, and the waitdone task waits for the completion of the command.

The main body of the test sends various commands to the controller and waits for their completion. It uses the `showdisk` task to display the data of the disk at the end of the test.

## Summary

This chapter described a complete Verilog model of a floppy disk subsystem. Although it is a simplified version of disk controller and disk drive, the model is useful for full system simulation because the intent is to debug the CPU section of the design. The final example demonstrated modeling techniques such as partitioning large models into smaller ones and using tasks, events, and functions to modularize the code and make it readable and maintainable.

The full model for the floppy disk subsystem, including the timing checker, the disk controller, the disk drive, and a top level test module, is listed at the end of the chapter.

## Source Listing

```verilog
'define f $finish;
'define STOP $stop

module disk_controller ( rd_, wr_, cs_, dack_, addr, dbus,
   irq, dma, moen_, dirc_, step_, byte_ctl_to_drv, tr00_,
   idx_, clk, byte_drv_to_ctl, mode, rst);

input rd_;
input wr_;
input cs_;
input dack_;
input addr;
inout [7:0] dbus;
output irq;
output dma;

output moen_;
output dirc_;
output step_;
output byte_ctl_to_drv;
input tr00_;
input idx_;
input clk;
input byte_drv_to_ctl;
output[1:0] mode;
input rst;

reg irq;
reg dma;
reg[7:0] dbus_reg;
wire[7:0] dbus = dbus_reg;

reg moen_, dirc_, step_;
reg[7:0] byte_ctl_to_drv;
wire tr00_, idx_, clk;
wire[7:0] byte_drv_to_ctl;

event do_read_byte;

'define IDLE    2'b00
'define COMMAND 2'b01
'define EXECUTE 2'b10

reg [1:0] mode;
wire idle_mode    = (mode == 'IDLE);
wire command_mode = (mode == 'COMMAND);
wire execute_mode = (mode == 'EXECUTE);

/* The main status register (main_stat_reg) bits */
parameter
   ST_COMPLETE = 0, /* The command was completed */
   ST_NOINDEX  = 1, /* Could not find the index mark */
   ST_ILLEGAL  = 2, /* Illegal command */
```

```verilog
      ST_CHECKSUM = 3, /* Bad checksum */
      ST_OVERRUN = 4, /* Overrun while sending to the host */
      ST_UNDERRUN = 5; /* Underrun while reading from the host */
reg[7:0] main_stat_reg;

parameter
    tDF = 1,
    tRI = 2;
parameter
    MAIN_STAT = 1,
    MAIN_DATA = 0;

reg[7:0] main_data_reg;
integer current_command; /* Index to the current command byte
                in command_array */
reg[7:0] command_array[1:2];
reg[7:0] checksum; /* The running checksum of the current
                sector */
integer current_track;

/* For debugging only*/
/*
initial begin
    $monitor ("dbus=%h, rd_=%b, wr_=%b, cs_=%b, irq=%b,
        dma=%b, dack_=%b, drv_ctl=%h, ctl_drv=%h time=%0d",
        dbus, rd_, wr_, cs_, irq, dma, dack_, byte_drv_to_ctl,
        byte_ctl_to_drv, $time); */
end
*/

parameter NCMDS = 6; /* Total number of possible commands */
integer new_command; /* The currently executing command */
parameter
    READ_DATA = 1,
    WRITE_DATA = 2,
    FORMAT_TRACK = 3,
    SEEK = 4,
    RECALIBRATE = 5,
    ILLEGAL_CMD = 0;

parameter
    USECS = 10, /* Time units per microsecond */
    MSECS = 10000, /* Time units per millisecond */
    SECTORS_PER_TRACK = 16,
    MAX_TRACK = 96,
    DATA_PER_SECTOR = 256,
    /* Data bytes + sector number + checksum */
    BYTES_PER_SECTOR = DATA_PER_SECTOR + 2,
    BYTES_PER_TRACK = BYTES_PER_SECTOR * SECTORS_PER_TRACK,
    BYTES_PER_DISK = BYTES_PER_TRACK * MAX_TRACK,
    TOTAL_SECTORS = SECTORS_PER_TRACK * MAX_TRACK,
    HALF_CYCLE = 2 * USECS,
    FULL_ROTATION = BYTES_PER_TRACK * 2 * HALF_CYCLE + 200;

integer cmd_bytes[0:NCMDS];/*command bytes for each command*/
reg [8*20:0] cmd_names [0:NCMDS];/* Command name strings */
```

```
integer cmd_code[0:NCMDS]; /* Command codes */


/*
The following three registers - SRT, HUT and HLT, hold the
delay timesfor the drive. The variable head_is_loaded
indicates whether the disk is rotating and is ready to
transfer data. The disk is shut off automatically after a
period specified in the HUT variable if there was no disk
access during this time. Once the disk is turned on, there is
a delay of HLT before it reaches full spinning speed and can
be accessed. SRT is the delay when the head is stepped from
one track to the next. The next two blocks (load_head_block
and unload_head_block) implement this automatic shutoff by
manipulating the variable "head_is_loaded" and the motor
signal "moen_".
*/

reg[3:0] SRT; /* Step Rate Time in mSecs increments */
reg[3:0] HUT; /* Head Unload Time in 16 mSecs increments */
reg[7:0] HLT; /* Head Load Time in 2 mSecs increments */
reg head_is_loaded; /* The current state of the head */
event load_head, unload_head;

always @load_head begin : load_head_block
   disable unload_head_block;
   moen_ = 0;
   #(HLT * MSECS * 2) head_is_loaded = 1;
end
always @unload_head begin : unload_head_block
   disable load_head_block;
   #(HUT * MSECS * 16) head_is_loaded = 0;
   moen_ = 1;
end

/*
The next two tasks initiate DMA transfers to and from the host
CPU respectively. They also check for overrun and underrun.
For debugging purpose, the two tasks stop when they detect
error. The real chip just set the status byte and continues.
*/

reg [7:0] byte_tofrom_host;
task send_byte_to_host;
begin
   if (dma) begin
      main_stat_reg[ST_OVERRUN] = 1;
      $display ("%m: byte overrun, byte = %h at time %0d",
         byte_tofrom_host, $time);
      'STOP;
   end
   dma = 1;
end
endtask

task get_byte_from_host;
begin
```

## Modeling a Floppy Disk Subsystem

```verilog
        dma = 1;
        dbus_reg = 8'hzz;
        @(negedge clk)
        if (dma) begin
            main_stat_reg[ST_UNDERRUN] = 1;
            $display ("%m: byte underrun at time %0d", $time);
            'STOP;
        end
    end
endtask

/*
Move the disk head by one track. First generate a step pulse,
and then delay by the amount specified by the SRT variable
*/

task one_step;
begin
    $display ("%m: at time %0d", $time);
    step_ = 0;
    #1 step_ = 1;
    #(SRT * MSECS) ;
end
endtask

/*
The next two tasks do the DMA transfer to and from the host CPU
respectively. The data byte is transfered between the data bus
(dbus) and an internal register "byte_tofrom_host".
*/

task dmaread;
begin
    /* $display ("%m: time = %0d", $time); */
    if (~execute_mode) begin
        $display ("%m: Spurious DMA read not in execute mode");
        'STOP;
    end
    dbus_reg = byte_tofrom_host;
    #tDF dbus_reg = 8'hZZ;
end
endtask

task dmawrite;
input[7:0] byte;
begin
    /* $display ("%m: byte = %h, time = %0d", byte, $time); */
    if (~execute_mode) begin
        $display ("%m: Spurious DMA read not in execute mode");
        'STOP;
    end
    byte_tofrom_host = dbus;
end
endtask

/*
```

```
The next few blocks implement the rst rd_ and wr_ inputs from
the host CPU. These are short blocks which call other tasks
to do the actual operation.
*/

always @(posedge rst) begin
   dbus_reg = 8'hzz;
   irq = 0;
   dma = 0;
   mode = 'IDLE;
   step_ = 1;
   dirc_ = 1;
   moen_ = 1;
   head_is_loaded = 0;
   disable load_head_block;
   disable unload_head_block;
end

always @(negedge rd_) begin
   if (dack_ == 1 && cs_ == 0) begin
      if (addr == MAIN_STAT)
         dbus_reg = main_stat_reg;
      else begin
         if (idle_mode)
            dbus_reg = main_data_reg;
         else begin
            $display ("Illegal read operation in mode %b",
               mode);
            'STOP;
         end
      end
      ->do_read_byte;
   end
   else if (~dack_)
      dmaread;
end

always @(posedge wr_) begin
   if (dack_ == 1 && cs_ == 0)
      programmed_write;
end

always @(posedge wr_) begin
   if (~dack_)
      dmawrite (dbus);
end

/*
The next two tasks do programmed read and write operations (as
opposed to DMA operations). They are initiated on the negative
edge of rd_ and positive edge of wr_ respectively. Depending
on the mode of the controller and on the address (addr) they
transfer the data between the internal registers and the data
bus, and possibly change the mode.
*/
```

```
task programmed_write;
begin
   if (addr != MAIN_DATA) begin
      $display ("Illegal register address: %b", addr);
      'STOP;
   end
   if (idle_mode || command_mode) begin
      if (idle_mode) begin
         current_command = 0;
         mode = 'COMMAND;
      end
      main_data_reg = dbus;
/*
The next piece of code writes one byte into the command_array
and increments the command byte counter. If a full command has
been written then go to execution mode and start executing the
command. When the command finishes executing, go back to idle
mode and generate an interrupt to the host.
*/
      current_command = current_command + 1;
      command_array[current_command] = dbus;
      if (current_command == 1) new_command = dbus;
      if (current_command == cmd_bytes[new_command]) begin
         mode = 'EXECUTE;
         main_stat_reg = 0;
         execute;
         main_stat_reg[ST_COMPLETE] = 1;
         mode = 'IDLE;
         irq = 1;
         ->unload_head;
      end
   end
   else begin
      $display ("Illegal write in unknown (%b) mode", mode);
      'STOP;
   end
end
endtask

/*
The next two blocks reset the irq and the dma outputs. The irq
is reset whenever a byte is read, and the dma is reset whenever
the dma is acknowledged by the host processor (negative edge
of dack_).
*/

always @do_read_byte begin
   @(posedge rd_)
   fork
      #tDF dbus_reg = 8'hZZ;
      #tRI irq = 0;
   join
end
always @(negedge dack_)
   #3 dma = 0;
```

```
/*
This is the main execution task. Some of the commands (SEEK
and RECALIBRATE) execute directly. The others (READ_DATA,
WRITE_DATA and FORMAT_TRACK) are delegated to another task:
read_write.
*/

task execute;
begin
   $display ("%m: Executing command %0d (%0s) at time %0d",
      new_command, cmd_names[new_command], $time);
   case (new_command)
   READ_DATA,
   WRITE_DATA,
   FORMAT_TRACK: begin
      read_write (new_command, command_array[2]);
   end
   SEEK: begin : seek_block
      integer diff, i;
      if (current_track > command_array[2]) begin
         diff = current_track - command_array[2];
         dirc_ = 1;
      end
      else begin
         diff = command_array[2] - current_track;
         dirc_ = 0;
      end
      for (i = 0; i < diff; i = i + 1)
         one_step;
      current_track = command_array[2];
   end
   RECALIBRATE: begin
      dirc_ = 1;
      while (tr00_)
         one_step;
      current_track = 0;
   end
   default: begin
      main_stat_reg[ST_ILLEGAL] = 1;
      $display ("Illegal command");
      'STOP;
   end
   endcase
end
endtask

/*
This is the main task for the commands READ_DATA, WRITE_DATA
and FORMAT_TRACK. It first makes sure that the head is loaded
and then waits for the index signal synchronization. The
FORMAT_TRACK command is executed directly. For the other
commands, this task just does search for the appropriate
sector, and then delegates the actual reading and writing of
data to the tasks read_data and write_data respectively.
*/

task read_write;
```

## Modeling a Floppy Disk Subsystem

```
input operation;
input [7:0] sec_byt;
integer operation;
integer sector, byte;
begin
   ->load_head;
   wait (head_is_loaded);
   $display ("%m: head is loaded at time %0d", $time);
   main_stat_reg[ST_NOINDEX] = 1;
   @(negedge idx_)
   main_stat_reg[ST_NOINDEX] = 0;
   $display ("Detected idx_ (command %0s) at time %0t",
      cmd_names [operation], $time);
   case (operation)
   FORMAT_TRACK: begin
      for (sector = 0;
         sector < SECTORS_PER_TRACK;
         sector = sector + 1) begin
         $display ("%m: Formatting sector %0d", sector);
         @(negedge clk) byte_ctl_to_drv = sector;
         checksum = sector;
         for (byte = 0;
            byte < DATA_PER_SECTOR;
            byte = byte + 1) begin
            /* $display ("Processing byte %0d", byte); */
            @(negedge clk) byte_ctl_to_drv = sec_byt;
            checksum = checksum + sec_byt;
         end
         @(negedge clk) byte_ctl_to_drv = checksum;
      end
      @(negedge clk) byte_ctl_to_drv = 8'hxx;
   end
   WRITE_DATA, READ_DATA: begin
      begin : search_sector
      for (sector = 0;
         sector < SECTORS_PER_TRACK;
         sector = sector + 1) begin
         $display ("Searching for sector = %0d", sector);
         @(posedge clk) if (byte_drv_to_ctl != sector) begin
            $display (
              "%m: Illegal sector number %0d instead of %0d",
              byte_drv_to_ctl, sector);
            'STOP;
            disable search_sector;
         end
         if (byte_drv_to_ctl == sec_byt) begin
            $display ("Found sector number (%0d)", sector);
            disable search_sector;
         end
         $disisplay(
          "%m:sector=%0d,byte_drv_to_ctl=%0d,sec_byt=%0d",
          sector, byte_drv_to_ctl, sec_byt);
         /* Still not the right sector. Skip the data */
         checksum = sector;
         for (byte = 0;
            byte < DATA_PER_SECTOR;
```

293

```verilog
                    byte = byte + 1) begin
                    /* $display ("Processing byte %0d", byte); */
                    @(posedge clk)
                    checksum = checksum + byte_drv_to_ctl;
                end
                @(posedge clk)
                if (byte_drv_to_ctl != checksum) begin
                    main_stat_reg[ST_CHECKSUM] = 1;
                    $display ("Bad checksum");
                    'STOP;
                end
            end
            $display ("Could not find the sector");
            'STOP;
            end /* search_sector block */
            checksum = sector;
            if (operation == WRITE_DATA)
                write_data;
            else
                read_data;
        end
        endcase
end
endtask

/*
The following two tasks write and read one sector's worth of
data to or from the host CPU. The data also include the
checksum byte.
*/

task write_data;
integer byte;
begin
    $display ("%m: Write_data to sector %0d", checksum);
    for (byte = 0;
         byte < DATA_PER_SECTOR;
         byte = byte + 1) begin
        get_byte_from_host;
        checksum = checksum + byte_tofrom_host;
        byte_ctl_to_drv = byte_tofrom_host;
    end
    get_byte_from_host;
    if (checksum != byte_tofrom_host) begin
        $display ("%m: Bad chacksum, %h, should be %h",
         byte_tofrom_host, checksum);
        'STOP;
    end
    byte_ctl_to_drv = byte_tofrom_host;
    @(negedge clk)
    byte_ctl_to_drv = 8'hxx;
end
endtask

task read_data;
integer byte;
```

# Modeling a Floppy Disk Subsystem

```verilog
      begin
         $display ("Read_data from sector %0d", checksum);
         byte_tofrom_host = checksum;
         send_byte_to_host;
         for (byte = 0;
              byte < DATA_PER_SECTOR;
              byte = byte + 1) begin
            @(posedge clk)
            checksum = checksum + byte_drv_to_ctl;
            byte_tofrom_host = byte_drv_to_ctl;
            /*$display("%m:Sending byte %0d(=%h) at host at %0d",
             byte, byte_drv_to_ctl, $time); */
            send_byte_to_host;
         end
         @(posedge clk)
         byte_tofrom_host = checksum;
         $display ("%m: Sending checksum byte %h to host at %0d",
            byte_drv_to_ctl, $time);
         send_byte_to_host;
         @(negedge dma) ;
         @(posedge rd_) ;
      end
   endtask

   /* Initialize the various command arrays */
   initial begin : set_arrays
      integer i;
      cmd_bytes[READ_DATA] = 2;
      cmd_bytes[WRITE_DATA] = 2;
      cmd_bytes[FORMAT_TRACK] = 2;
      cmd_bytes[SEEK] = 2;
      cmd_bytes[RECALIBRATE] = 1;
      cmd_bytes[ILLEGAL_CMD] = 0;

      cmd_names[READ_DATA] = "READ_DATA";
      cmd_names[WRITE_DATA] = "WRITE_DATA";
      cmd_names[FORMAT_TRACK] = "FORMAT_TRACK";
      cmd_names[SEEK] = "SEEK";
      cmd_names[RECALIBRATE] = "RECALIBRATE";
      cmd_names[ILLEGAL_CMD] = "ILLEGAL_CMD";

      for (i = 0; i <= NCMDS; i = i + 1)
         cmd_code[i] = ILLEGAL_CMD;
      cmd_code['b00110] = READ_DATA;
      cmd_code['b00101] = WRITE_DATA;
      cmd_code['b01101] = FORMAT_TRACK;
      cmd_code['b01111] = SEEK;
      cmd_code['b00111] = RECALIBRATE;
      SRT = 2;
      HUT = 3;
      HLT = 5;
   end
endmodule /* disk_controller */

/* =========================================================== */
```

```
This is the top module that is used to test the fdc module.
It has one instance of fdc and behavioral code to generate
test vectors on the inputs of the fdc and to observe the output
of the fdc.
*/

module test_fdc;
reg [7:0] dbus_reg;
wire [7:0] dbus = dbus_reg;
reg rst, rd_, wr_, cs_, dack_;
reg addr;
reg debug;
reg receiving;

parameter BUFFSIZE = 1024;
reg [7:0] buff[0:BUFFSIZE-1]; /* Hold one sector worth of data
*/
integer nbuff; /* The number of data in the buffer */

parameter
    MAIN_STAT = 1,
    MAIN_DATA = 2;
parameter
    READ_DATA = 1,
    WRITE_DATA = 2,
    FORMAT_TRACK = 3,
    SEEK = 4,
    RECALIBRATE = 5,
    ILLEGAL_CMD = 0;


/* Wires connecting the controller to the drive */
wire moen_, dirc_, step_, tr00_, idx_, clk;
wire [7:0] byte_ctl_to_drv, byte_drv_to_ctl;

wire [1:0] mode;

/* Check input signal validity */
checktiming ct (rd_, wr_, cs_, dack_, addr, dbus, mode);

/* Do CPU interface */
disk_controller fdc (rd_, wr_, cs_, dack_, addr, dbus, irq,
dma,
    moen_, dirc_, step_, byte_ctl_to_drv, tr00_, idx_,
    clk, byte_drv_to_ctl, mode, rst);

/* Do floppy drive interface */
disk_drive fdd (moen_, dirc_, step_, byte_ctl_to_drv, rst,
    tr00_, idx_, clk, byte_drv_to_ctl);

/* Process DMA request by the controller */
always @(posedge dma) begin
    if (receiving) begin
        #1 dack_ = 0;
        #2 rd_ = 0;
```

## Modeling a Floppy Disk Subsystem

```verilog
            #1 buff[nbuff] = dbus;
            nbuff = nbuff + 1;
            #3 rd_ = 1;
            #1 dack_ = 1;
        end
        else begin
            /* $display ("%m: sending buff[%0d] = %h", nbuff,
buff[nbuff]); */
            #1 dack_ = 0;
            dbus_reg = buff[nbuff];
            #2 wr_ = 0;
            nbuff = nbuff + 1;
            #3 wr_ = 1;
            #1 dbus_reg = 8'hzz;
            dack_ = 1;
        end
end

/* Initialize the buffer for reading from the controller */
task initbuff;
integer i;
begin
    for (i = 0; i < BUFFSIZE; i = i + 1)
        buff[i] = 8'hxx;
    nbuff = 0;
end
endtask

/* Initialize the buffer for writing to the controller */
task initbuffwrite;
integer i;
begin
    for (i = 0; i < BUFFSIZE; i = i + 1)
        buff[i] = i;
    nbuff = 0;
end
endtask

/* A task and a macro to show the contents of the buffer */
'define sb showbuff; #0 $stop; .
task showbuff;
integer i, j, linesize;
reg nonx;
begin
    for (i = 0; i < BUFFSIZE; i = i + linesize) begin
        nonx = 0;
        linesize = 20;
        if (i + linesize > BUFFSIZE) linesize = BUFFSIZE - i;
        for (j = i; j < i + linesize; j = j + 1)
            if (buff[j] !== 8'hxx) nonx = 1;
        if (nonx) begin
            $write ("\nbuff[%d]: ", i);
            for (j = i; j < i + linesize; j = j + 1)
                $write (" %h", buff[j]);
        end
    end
```

```
         $display ("");
      end
   endtask

   /* Reset the controller */
   task dorst;
   begin
      rst = 1;
      #100 rst = 0;
   end
   endtask

   /*
   The next two tasks send and receive one byte from the
   controller in programmed mode.
   */

   task sndbyte;
   input [7:0] byte;
   input add;
   begin
      if (debug)
         $display ("%m: byte = %h, add = %b, time = %0d",
            byte, add, $time);
      cs_ = 0;
      addr = add;
      #2 wr_ = 0;
      #1 dbus_reg = byte;
      #10 wr_ = 1;
      #2 cs_ = 1;
      dbus_reg = 8'hzz;
   end
   endtask

   task rcvbyte;
   output [7:0] byte;
   input add;
   begin
      $display ("%m: address = %b", add);
      cs_ = 0;
      addr = add;
      #2 rd_ = 0;
      #1 byte = dbus;
      #3 rd_ = 1;
      #1 cs_ = 1;
      $display ("Received byte = %h", byte);
   end
   endtask

   /*
   The next two tasks send commands of length 1 and 2
   respectively to the controller. They call the task sndbyte
   with address MAIN_DATA.
   */
```

```
task sndcmd1;
input[7:0] cmd;
begin
    $display ("%m: cmd = %h", cmd);
    if (irq) begin
        $display ("%m: irq is high, cannot initiate a
command");
        $stop;
    end
    sndbyte (cmd, MAIN_DATA);
end
endtask

task sndcmd2;
input [7:0] cmd;
input [7:0] byte1;
begin
    $display ("%m: cmd = %h, byte1 = %h",
        cmd, byte1);
    if (irq) begin
        $display ("%m: irq is high, cannot initiate a
command");
        $stop;
    end
    sndbyte (cmd, MAIN_DATA);
    sndbyte (byte1, MAIN_DATA);
end
endtask

task waitdone;
reg [7:0] status;
begin
    wait (irq);
    #1
    rcvbyte (status, MAIN_STAT);
    if (status != 1) begin
        $display ("%m: Status = %b (instead of 00000001 at time
%0d", status, $time);
        $stop;
    end
    wait (~irq);
end
endtask

/*
The actual test is composed of a string of lower level tasks.
The following actions take place:
1. Reset the controller.
2. Send the RECALIBRATE command to retract head to track 0.
3. Send a SEEK comand to move the head to track 3.
4. Send two FORMAT commands for the current track
    (one of the commands is redundant).
5. Send a READ_DATA command to read the formatted track
    sector.
6. Send a WRITE_DATA command to write on one of the sectors.
```

```verilog
*/
initial begin
$monitor ("irq = %b at time %0d", irq, $time);
    debug = 0;
    rst = 0;
    rd_ = 1;
    wr_ = 1;
    cs_ = 1;
    dack_ = 1;
    #1 dorst;
    sndcmd1 (RECALIBRATE);
    waitdone;
    sndcmd2 (SEEK, 3);
    waitdone;
    sndcmd2 (FORMAT_TRACK, 8'h33);
    waitdone;
    initbuff;
    receiving = 1;
    sndcmd2 (READ_DATA, 1);
    waitdone;
    showbuff;
    receiving = 0;
    initbuffwrite;
    sndcmd2 (WRITE_DATA, 0);
    waitdone;
    $display ("Test complete");
    showbuff;
    test_fdc.fdd.showdisk;
    $stop;
end

endmodule /* test_fdc */


/* ====================================================== */
module checktiming (rd_, wr_, cs_, dack_, addr, dbus, mode);
input rd_;
input wr_;
input cs_;
input dack_;
input addr;
input[7:0] dbus;
input[1:0] mode;
wire execute_mode = (mode == 'EXECUTE);
wire idle_mode = (mode == 'IDLE);
wire command_mode = (mode == 'COMMAND);

/*
The following parameters are the various timing constraints as
specified in the data sheet and shown in the waveform diagram.
For simplification, all the delays are assumed to be 0 or 1
time units. The actual delays are given in the comments, in
nanoseconds. The variables that follow remember the time at
which the various signals have changed. These times are used
when the timing constraints are checked.
```

```
*/

parameter
    tARmin = 1, /* 10 */
    tRRmin = 1, /* 80 */
    tRAmin = 0,
    tAWmin = 1, /* 10 */
    tWDmin = 1, /* 70 */
    tWAmin = 0;
time rd_high,rd_low,wr_high,cs_low,dack_high,addrchanged;

wire programmed_io = (idle_mode || command_mode) && !cs_;

/*
The folloing three short tasks are used for improved
readability. The first task just prints a message and stops.
The second task checks for minimum delay violation and stops
if one is detected, and the third one checks for maximum delay
violation. The tasks call the macro 'STOP which is defined as
$stop, however, the macro can be modified to be null, so that
timing violations are reported but simulation continues.
*/

task illegal_signal;
input signame;
input sigval;
reg [8*20:0] signame;
reg sigval;
begin
    $display ("%m: Illegal signal %s = %b at time %0d",
        signame, sigval, $time);
    'STOP;
end
endtask

task checkmintiming;
input message;
input tdiff;
input prevtime;
reg [8*20:0] message;
integer tdiff;
time prevtime;
begin
    if ($time - prevtime < tdiff) begin
        $display ("%m: !! Timing violation: %s, %0d - %0d < %0d",
            message, $time, prevtime, tdiff);
        'STOP;
    end
end
endtask

task checkmaxtiming;
input message;
input tdiff;
input prevtime;
```

```
        reg [8*20:0] message;
        integer tdiff;
        time prevtime;
        begin
            if ($time - prevtime > tdiff) begin
                $display ("%m: Timing violation: %s, %0d - %0d > %0d",
                    message, $time, prevtime, tdiff);
                'STOP;
            end
        end
endtask

initial begin
    rd_high = 0;
    rd_low = 0;
    wr_high = 0;
    cs_low = 0;
    dack_high = 0;
    addrchanged = 0;
end

/*
The following few blocks just record the time at which the
various bus signal have been modified. These times are later
used to do timing checks.
*/

always @rd_
    if (rd_ == 1)
        rd_high = $time;
    else if (rd_ == 0)
        rd_low = $time;
    else if (programmed_io) illegal_signal ("rd_", rd_);

always @wr_
    if (wr_ == 1)
        wr_high = $time;
    else if (programmed_io && wr_ != 0)
        illegal_signal ("wr_", wr_);

always @cs_
    if (cs_ == 0)
        cs_low = $time;
    else if (cs_ != 1)
        illegal_signal ("cs_", cs_);

always @dack_
    if (dack_ == 1)
        dack_high = $time;

always @addr addrchanged = $time;

/*
The rest of the blocks do the actual timing checks by calling
the previously defiined tasks.
*/
```

```
always @(negedge rd_) if (programmed_io) begin
   checkmintiming ("tAR to cs_", tARmin, cs_low);
   checkmintiming ("tAR to addr", tARmin, addrchanged);
   checkmintiming ("tAR to dack_", tARmin, dack_high);
end

always @(negedge wr_) if (programmed_io) begin
   checkmintiming ("tAW to cs_", tAWmin, cs_low);
   checkmintiming ("tAW to addr", tAWmin, addrchanged);
   checkmintiming ("tAW to dack_", tAWmin, dack_high);
end

always @(posedge rd_) if (programmed_io) begin
   checkmintiming ("tRR", tRRmin, rd_low);
end

always @dbus if (programmed_io) begin
   checkmintiming ("tWD", tWDmin, wr_high);
end

always @(posedge cs_ or negedge dack_ or addr) begin
   checkmintiming ("tRAmin", tRAmin, rd_high);
   checkmintiming ("tWAmin", tWAmin, wr_high);
end
endmodule
```

/* ================================================== */
/*
The disk drive model simulates the physical drive unit,
including the actual floppy disk and the information that it
contains. The following signals communicate between the drive
and the controller:

moen_ -- Motor enable (low active). Input. When active, then
      after some delay the disk starts rotating and sending
      idx_ pulses.
dirc_ -- Direction (low active). Input. When active, then each
      step pulse moves the head outside, otherwise, each step
      pulse moves the head back inside.
step_ -- Step (low active). Input. Each pulse moves the head
      one
      cylinder in a direction specified by dirc_.
byte_in -- Byte input. Input. Input byte from the controller.
tr00_ -- Track zero (low active). Output. When active, it
      indicates that the head is retracted to track zero.
idx_ -- Index (low active). Output. Every revolution of the
      diskette, the drive sends a pulse to the controller.
      This indicates the beginning of the track
      (first sector).
clk -- Clock. Output. Used to synchronize the drive with the
      controller. In a real floppy drive the clock is derived
      from the data.
byte_out -- Byte output. Output. Output byte to the
      controller.

```verilog
   rst -- Reset
*/

module disk_drive (moen_, dirc_, step_, byte_in, rst,
    tr00_, idx_, clk, byte_out);

input moen_;
input dirc_;
input step_;
input [7:0] byte_in;
input rst;

output tr00_;
output idx_;
output clk;
output [7:0] byte_out;

parameter
   USECS = 10, /* Time units per microsecond */
   MSECS = 10000, /* Time units per millisecond */
   SECTORS_PER_TRACK = 16,
   MAX_TRACK = 96,
   DATA_PER_SECTOR = 256,
   BYTES_PER_SECTOR = DATA_PER_SECTOR + 2, /* Data bytes +
sector number + checksum */
   BYTES_PER_TRACK = BYTES_PER_SECTOR * SECTORS_PER_TRACK,
   BYTES_PER_DISK = BYTES_PER_TRACK * MAX_TRACK,
   TOTAL_SECTORS = SECTORS_PER_TRACK * MAX_TRACK,
   HALF_CYCLE = 2 * USECS,
   FULL_ROTATION = BYTES_PER_TRACK * 2 * HALF_CYCLE + 200;

parameter RAMPTIME = 10 * MSECS; /* uSecs for full speed */

event do_idx;
integer byte_index;
reg [7:0] checksum;
reg idx_;
reg clk;
reg [7:0] byte_out;
reg [MAX_TRACK-1:0] formatted;/*One bit per track indicating
             /* whether the track s formatted or not */
reg rotating; /* Indicates whether the disk is rotating */
reg [7:0] diskmem [0:BYTES_PER_DISK]; /* Holds the disk data*/
integer current_track;
wire tr00_ = (current_track != 0);

task init;
integer i;
begin
   $display ("The total number of bytes is %0d",
BYTES_PER_DISK);
   $display ("Full rotation = %0d uSecs", FULL_ROTATION);
   for (i = 0; i < MAX_TRACK; i = i + 1)
      formatted[i] = 0;
   idx_ = 1;
   current_track = 0;
```

```verilog
        rotating = 0;
    end
endtask

initial init;

/*
The following block implements the head movement. The variable
current_track is the current track. Each (negative)
pulse on the step_ input increments or decrements
current_track,
depending on the value of the direction input (dirc_).
*/

always @(posedge step_) begin
    if (dirc_ == 1 && current_track > 0)
        current_track = current_track - 1;
    if (dirc_ == 0 && current_track < MAX_TRACK - 1)
        current_track = current_track + 1;
end

/*
The following two blocks generate the index signal of the
disk. Whenever the motor is on, the disk rotates and generates
an index signal once per rotation. Whenever the disk motor
shuts off (moen_ == 1) and then turns on, there is a RAMPTIME
period before the signal "rotating" turns on.
*/

always begin
    #FULL_ROTATION
    ->do_idx;
    if (rotating) begin
        idx_ = 0;
        #1 idx_ = 1;
    end
end

always @moen_ begin
    rotating = 0;
    idx_ = 1;
    if (~moen_) begin
        #RAMPTIME
        if (~moen_) rotating = 1;
    end
end

/*
The next two blocks implement the clock generation. The clock
starts operating when the disk rotates. It is also
synchronized to the index signal every full rotation.
*/

always @(negedge rotating) disable clock_gen;

always begin : clock_gen
```

```verilog
      integer cycle;
         wait (rotating);
         cycle = 0;
         forever begin
            clk = 1;
            #HALF_CYCLE clk = 0;
            #HALF_CYCLE clk = 1;
            cycle = cycle + 1;
         end
   end

   /*
   While the disk is rotating, the following generate_byte block
   checks if the track is formatted. If so, then generate_byte
   generates clocks and data, one per byte. It does it in three
   sections, sector head, sector data and sector tail. Sector
   head just sends
   */

   always @do_idx begin : generate_byte
      integer sector;
      disable clock_gen;
      if (~rotating) begin
         byte_out = 8'hxx;
         disable generate_byte;
      end
      byte_index = current_track * BYTES_PER_TRACK;
      for (sector = 0; sector < SECTORS_PER_TRACK;
                 sector = sector + 1) begin
         sector_head (sector);
         sector_data;
         sector_tail;
      end
   end

   always @(posedge rst) begin
      $display ("%m: reset");
      disable generate_byte;
      disable clock_gen;
      disable sector_head;
      disable sector_data;
      disable sector_tail;
   end

   always @(negedge rotating) disable generate_byte;

   /*
   The next few short tasks are used for transferring data bytes
   between the disk drive and the controller. The bytes in the
   drive are organized in one big array. A global variable
   byte_index holds the index to the next byte in the array which
   should be read or written, and this index is incremented after
   every transfer.
   */
```

```
task sendbyte;
begin
    /* $display ("%m: byte_index = %0d", byte_index); */
    byte_out = diskmem[byte_index];
    byte_index = byte_index + 1;
end
endtask

task sendbyte_x;
begin
    /* $display ("%m: byte_index = %0d", byte_index); */
    byte_out = 8'hxx;
    byte_index = byte_index + 1;
end
endtask

task getbyte;
begin
    /* $display ("%m: byte_index-1 = %0d, byte_in = %h",
        byte_index-1, byte_in); */
    diskmem[byte_index - 1] = byte_in;
end
endtask

/*
The following three tasks do the actual transfer of the sector
head, sector data and sector tail respectively. The sector
head is one byte containing the ordinal number of the sector
on the track (starting from zero). The sector data contains
DATA_PER_SECTOR contiguous bytes, and the sector tail is a one
byte checksum, which is calculated as the sum of all the data
bytes and the sector number, ignoring overflow. The disk drive
communicates with the controller by two 8 bit buses, one
transfers bytes from the drive to the controller and the other
transfers bytes from the controller to the drive. Data are
presented on these buses on the negative edge of the clock
(clk) and are sampled
on the positive edge of the clock. The drive always sends the
sector information on the outgoing bus. It updates the disk
data only if the incoming byte is not 8'hxx.
*/

task sector_head; input sector; integer sector; begin
    /* $display ("%m: sector = %0d", sector); */
    checksum = diskmem[byte_index];
    @(negedge clk)
    if (formatted[current_track]) sendbyte;
    else sendbyte_x;
    @(posedge clk) if (byte_in !== 8'hxx) begin
        getbyte;
        formatted[current_track] = 1;
    end
end
endtask

task sector_data;
```

```verilog
        integer i;
        begin
            for (i = 0; i < DATA_PER_SECTOR; i = i + 1) begin
                @(negedge clk) sendbyte;
                @(posedge clk) if (byte_in !== 8'hxx)
                    getbyte;
                checksum = checksum + diskmem[byte_index-1];
            end
        end
    endtask

    task sector_tail;
        begin
            /* $display ("%m"); */
            @(negedge clk) sendbyte;
            @(posedge clk) if (byte_in !== 8'hxx)
                getbyte;
            if (checksum != diskmem[byte_index - 1]) begin
                $display ("%m: Bad checksum %h, should be %h",
                    diskmem[byte_index-1], checksum);
                $stop;
                diskmem[byte_index-1] = checksum;
            end
        end
    endtask

    /*
    The following task displays the disk information for debugging
    purposes. The macro sd is a short notation for invoking this
    task. This macro can be called from any $scope, since it has
    the full hierarchical name of the task
    */

    `define sd test_fdc.f.di.showdisk; #0 $stop; .
    task showdisk;
    integer i, j, k, index;
    begin
        for (i = 0; i < MAX_TRACK; i = i + 1) begin : track_block
            if (~formatted[i]) disable track_block;
            index = i * BYTES_PER_TRACK;
            $display ("Track = %0d, index = %0d", i, index);
            for (j = 0; j < SECTORS_PER_TRACK; j = j + 1) begin
                $write ("Sector %0d: ", j);
                for (k = 0; k < BYTES_PER_SECTOR; k = k + 1) begin
                    $write (" %h", diskmem[index]);
                    index = index + 1;
                end
                $display (";");
            end
        end
    end
    endtask

endmodule /* disk_drive */
```

# CHAPTER 9

# Useful Modeling and Debugging Techniques

Learning to design and simulate in Verilog is more than just learning the syntax and semantics of the language. As in every learning process, the best way to learn is by doing. As you start using the language, you will develop your own style of design, and you will discover techniques for modeling in Verilog. In this chapter we present some tips and techniques that we hope will help you in developing your own techniques on the way to mastering Verilog.

## Bidirectional Ports

A bidirectional port operates as either an input or an output, depending on a control variable. Typically, a data bus is shared among the CPU, the memory system, and other peripheral devices. The bus connection in each module is modeled by a bidirectional port. At any time, only one device can write to the bus and the rest of the devices can only read.

A port in Verilog can be designated as `input`, `output` or `inout`. In a Verilog behavioral model, the `input` ports are wires and the `output` ports are registers. A bidirectional (`inout`) port can be specified as a wire which accepts its value from a "shadow register" through a continuous assignment. When the port is enabled as an output, you set the value of the shadow register to the desired output value. When the port is disabled for output and should operate as an input, simply set the shadow register to Z. This enables the other signal input to the port.

Figure 9.1 shows a complete example of modeling a bidirectional port. Figure 9.2 shows a block diagram of a CPU with memory. The data bus is bidirectional, and depending on whether the CPU reads from memory or writes to memory, the data port operates as an input or an output. The signal waveforms are shown in Figure 9.3 and Figure 9.4. When the memory senses the `read` signal, it puts the appropriate data on the data bus; the `data` port is an output from the memory and is an input to the CPU. When the memory senses a negative transition on the `write` signal, it stores the value from the data bus to the appropriate

```
module top;
wire [31:0] address;
wire [15:0] data;
wire read, write;

   cpu u1 (data, address, read, write);
   mem u2 (data, address, read, write);

endmodule

module cpu (data, address, read, write);
output [31:0] address;
inout  [15:0] data;
output read, write;

reg [31:0] address;
reg [15:0] data_reg;
wire [15:0] data = data_reg;
reg read, write;

reg [15:0] somedata;

   initial
      data_reg = 16'hzzzz;

task writetomem;
input [31:0] taddress;
input [15:0] tdata;
```
**continued**

**Figure 9.1** Code for modeling a bidirectional port

## Useful Modeling and Debugging Techniques

```
 begin
   data_reg = tdata;
   #1 write = 1;
   #20 write = 0;
   #1 data_reg = 16'hzzzz;
end

endtask

task readfrommem;
input  [31:0] taddress;
output [15:0] tdata;

begin
   read = 1;
   #20 tdata = data;
   #1 read = 0;
end

endtask

   initial begin
       ......
       writetomem (32'h100, 16'h5555);
       ......
       readfrommem (32'h104, somedata);
       ......
   end
   ...
endmodule

module mem (data, address, read, write);
input [31:0] address;
inout [15:0] data;
input read, write;

wire [31:0] address;
reg  [15:0] data_reg;
wire [15:0] data = data_reg;
wire read, write;

reg  [15:0] memarr [0:1023];

always @(posedge write) begin
   @(negedge write)
   memarr[address] = data;
end

always @(posedge read) begin
   data_reg = memarr[address];
   @(negedge read)
   data_reg = 32'hzzzzzzzz;
end

endmodule
```

**Figure 9.1** Code for modeling a bidirectional port(continued)

# Digital Design and Synthesis with Verilog HDL

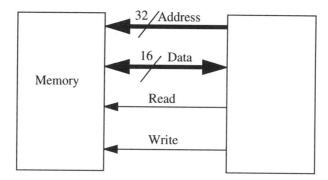

**Figure 9.2** CPU block diagram

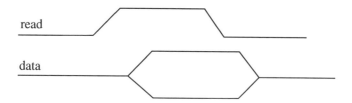

**Figure 9.3** Waveform timing diagram for a read cycle

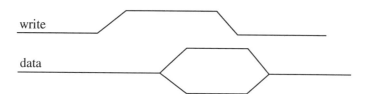

**Figure 9.4** Waveform timing diagram for a write cycle

address; the `data` port is an output from the CPU and is an input to the memory.

## Bus Transactions in a Pipeline Architecture

It is common in a pipeline architecture for bus transactions to extend over multiple cycles. Usually such transactions are interleaved to

## Useful Modeling and Debugging Techniques

maintain the throughput of one transaction per cycle. Consider a CPU/memory system, in which the address is transferred to the address bus in two cycles. The address bus has four fields as follows:

| 33 | 32      | 17 16 | 15       0 |
|----|---------|-------|------------|
| W  | MSB[15:0] | V   | LSB[15:0]  |

When bit 33, the W bit, is 1, the current transaction is a write transaction; otherwise it is a read transaction. Bits 32-17 are the 16 most significant address bits of the current transaction. Bit 16, the V bit, if 1, indicates that the previous transaction was valid; otherwise abort the

```
W[1]   MSB[1]   V[0]   LSB[0]
W[2]   MSB[2]   V[1]   LSB[1]
W[3]   MSB[3]   V[2]   LSB[2]
W[4]   MSB[4]   V[3]   LSB[3]
W[5]   MSB[5]   V[4]   LSB[4]
W[6]   MSB[6]   V[5]   LSB[5]
```

**Figure 9.5** Address bus transactions

previous transaction. Bits 15-0 are the least significant bits of the previous transaction. Figure 9.5 depicts a snapshot of the address bus where the number in brackets indicates the transaction number.

It is desirable to write a task that accepts as input the four fields of an address bus transaction (W, MSB, V, and LSB) and generates the transaction. Such encapsulation enables you to issue transactions from different locations in the code. More importantly, you can issue transactions interactively from the terminal. A straightforward approach of modeling such a task is depicted in Figure 9.6. Assume that addressbus[33:0] was declared previously in the module.

But the task as written in this figure is not sufficient. If two consecutive bus transactions are needed, the following statements will not work:

```
@clock #1 addrtrans ('x100, 1, 1);
@clock #1 addrtrans ('x102, 1, 0);
```

313

```
task addrtrans;
input [31:0] address;
input valid;
input write;

begin
   @clock
   addressbus[33:17] = { write, address[31:16] };
   @clock
   addressbus[16:0] = { valid, address[15:0] };
end

endtask
```

**Figure 9.6** Modeling bus transactions

```
......
reg   [15:0] msb_address, lsb_address;
reg   vld, wrt;
event do_trans;

task addrtrans;
input [31:0] address;
input valid;
input write;
begin
   msb_address = address[31:16];
   lsb_address = address[15:0];
   vld = valid;
   wrt = write;
   ->do_trans;
end

endtask

   always @do_trans begin : trans_block
      @clock
      addressbus[33:17] = { wrt, msb_address };
      @clock
      addressbus[16:0] = { vld, lsb_address };
   end
```

**Figure 9.7** An attempt to solve the multiple-cycle transactions

# Useful Modeling and Debugging Techniques

Since each execution of the task takes two clock cycles, the second call to the `addrtrans` task occurs two clock cycles after the first one. In other words, there is no way to execute two transactions on two consecutive clock cycles; thus the effect of the pipeline can not be simulated. One way to try to correct this problem is to use events as shown in Figure 9.7.

This solution has an advantage over the Figure 9.6 model in that the addrtran task takes no simulation time to execute. In effect, the task does not perform the transaction, but just triggers it. The actual transaction is being done by `trans_block`. However, each loop takes two cycles to execute, so that the code in Figure 9.7 still does not work. The first transaction starts execution, and one clock into the transaction the `do_trans` event is triggered again; but this event is lost because the

```
......
reg    [15:0] msb_address, lsb_address, lsb_address1;
reg    vld, wrt, vld1;
event do_trans_cycle1, do_trans_cycle2;

task addrtrans;
input [31:0] address;
input valid;
input write;

begin
   msb_address = address[31:16];
   lsb_address = address[15:0];
   vld = valid;
   wrt = write;
   ->do_trans_cycle1;
end

endtask

   always @do_trans_cycle1 begin
      @clock
      addressbus[33:17] = { wrt, msb_address };
      vld1 = vld;
      lsb_address1 = lsb_address;
      ->do_trans_cycle2;
   end

   always @do_trans_cycle2 begin
      @clock
      addressbus[16:0] = { vld1, lsb_address1 };
   end
```

**Figure 9.8** A correct way to model multiple bus transactions

always loop did not complete and is not waiting for the event to be triggered.

To solve this problem, we use two always loops and two events. As before, the task just triggers an event that enables an always loop. This time, however, the loop is only one clock long, and triggers another event that enables the second always loop (Figure 9.8).

The extension to longer transactions of three or more cycles is straightforward. Use a sequence of always loops, each of them lasting one cycle and triggering an event that enables the next cycle. Notice that you also need temporary global variables that pass the pipeline values from cycle to cycle. In the example above these variables are msb_address, lsb_address, lsb_address1, vld, wrt, and vld1.

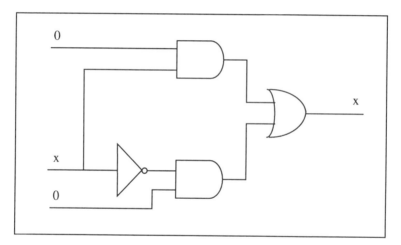

**Figure 9.9** Unknown inputs in combinational networks

## Combinational Blocks with Unknown Inputs

The outputs of a combinational block might have known values even when the inputs are unknown. The simplest example of such a case is an AND gate that has one input X and another input 0. In such a case, Verilog recognizes that the output of the gate must be 0. A slightly more

# Useful Modeling and Debugging Techniques

complicated example is a 2-to-1 selector in which both inputs are 0 and the select input is X (Figure 9.9). Even though the output is 0 no matter what the select input is, Verilog does not recognize it and instead propagates an X to the output. One way to get the correct result is to design a selector UDP that exhibits the desired behavior under all input conditions. This might not be feasible for more complicated circuits.

In Verilog, an X value on a signal can have two different causes. One cause is that two different sources try to drive the same net (using the same strength) to two different logical values. This is usually a design

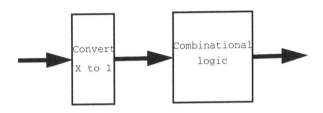

**Figure 9.10** Converting unknown inputs to known inputs

```
module xto0 (out, in);
output out;
input  in;

   wire out = (in == 1) ? 1 : 0;

endmodule
```

**Figure 9.11** A device for converting unknown signals

error. The other possible cause for an X is that a signal has not been initialized. Sometimes an uninitialized signal also indicates a design error, but X's during the first cycles of initialization are legitimate and a designer can exploit the cases where the output is independent of the X inputs in order to optimize the design.

Suppose that a state machine is described in Espresso (UC Berkeley) format, and that some entries in the state table have remained unspecified to optimize design. In order to force the output to a known value, you may have to force all the inputs to some arbitrary (but known) values. You can insert a buffer device that converts X to 1 (or 0) as shown in Figure 9.10. The implementation of such a device is given in Figure 9.11.

## Large Memory as a Table Lookup

When modeling a large memory subsystem, you might encounter degradation in simulation performance. Verilog itself does not place any limit on the size of a memory array, but if the memory has an address space of four gigabytes, chances are that the operating system will not be able to allocate space for such a large array. One way to solve the problem is to use a table lookup technique (or associative memory) to simulate the physical memory.

Our first solution assumes that memory access is highly localized; namely, that at each point in time only a small fraction of the memory is used. For example, let's assume that only sixteen memory locations are used at any one time. Two arrays are provided: one for memory addresses and one for memory contents. Whenever we access a memory location, all memory addresses are scanned to see if a named address is in use. If the address is in use, the corresponding memory contents array location is used for reading or writing. If this memory location is not being used, the address and contents are inserted into the two arrays. In the latter case, another memory location may have to be replaced, perhaps using the least-recently-used (LRU) algorithm.

If we attempt to read a nonexisting memory location, the program prints an error message and stops. Figure 9.12 shows the code for the memory module. For simplicity, this model does not use the LRU algorithm to release array locations; instead, a rotating counter selects the next location to be released.

The solution in Figure 9.12 is effective only for very localized memory access. If our program uses a large memory in a random way, this scheme can be very slow. Every time memory is accessed, the complete array (16 here) is scanned for the required address. If the array size must be increased to say, 4096, simulation may run too slowly. Here we can borrow an algorithm from software engineering, called "hash table access," to reduce search time.

Assume that we will access no more than 4096 memory locations, scattered randomly between 0 and 1 megabytes. In this solution, as in the previous one, we have two arrays: one for addresses and one for contents. Instead of scanning all the addresses, a hash function converts the real address into an index to the tables. The array size is 5041 (a prime number about 1.2 times larger than the total number of entries), and the hash function is the remainder of the real address when divided by the hash table size. Figure 9.13 shows the hash function for accessing a memory location. Figure 9.14 and Figure 9.15

```
module memory (address, data, rd, wr);
input [31:0] address;
inout [15:0] data;
input rd, wr;

parameter MEMSIZE = 16;

wire [31:0] address;
reg  [15:0] data_reg;
wire [15:0] data = data_reg;
wire rd, wr;

reg [31:0] address_array[0:MEMSIZE-1];
reg [15:0] data_array[0:MEMSIZE-1];

integer lastwritten;

   initial
      lastwritten = 0;

   always @(posedge wr) begin : write_cycle
      integer i;
      #1 // Set up time
      /*
      First see if this address exists in the address_array
       if yes, then just replace the data in this
       address; otherwise
       replace the data in the location specified by
       lastwritten.
      */
      for (i = 0; i <= 15; i = i + 1)
         if (address_array[i] == address)
            lastwritten = i;
      address_array[lastwritten] = address;
      data_array[lastwritten] = data;
      lastwritten = (lastwritten + 1) % 16;
   end // write_cycle                    continued
```

**Figure 9.12** Associative array for simulating large memory

```
always @(posedge rd) begin : read_cycle
    integer i;
    integer found;
    found = 0;
    /*
    See if the address exists in the address_array. If
    yes, then fetch the data; otherwise print an
    error message and stop the simulation. Wait for
    the negative edge of read and tristate the data bus.
    */
    for (i = 0; i <= 15; i = i + 1)
        if (address_array[i] == address) begin
            found = 1;
            data_reg = data_array[i];
        end
    if (found == 0) begin
        $display (
        "Trying to read nonexisting data fromaddress %h",
            address);
        $stop;
    end
    @(negedge rd)
    data_reg = 16'hzzzz;
end // read_cycle

endmodule
```

**Figure 9.12** Associative array for simulating large memory(continued)

```
.....
'define HASHSIZE 5041
reg  [31:0] address_tab [0:'HASHSIZE-1];
reg  [31:0] data_tab [0:'HASHSIZE-1];

initial begin : intitialize_hashtab
    integer i;
    for (i = 0; i < 'HASHSIZE; i = i + 1)
        address_tab[i] = 'hxxxxxxxx;
    end
end

function integer hash;
input [31:0] address;

    hash = address % 'HASHSIZE;

endfunction
```

**Figure 9.13** Hash function and hash table initialization

```
function [31:0] readdata;
input [31:0] address;
integer index, i;

begin
   index = hash (address);
   if (address_tab[index] == address)
      readdata = data_tab[index];
   else begin : search_address
      // The first hashing did not produce the right address
      // Start scanning the table for the address
      for (i = (index + 1) % 'HASHSIZE;
           i != index;
           i = (i + 1) % 'HASHSIZE) begin
         // Due to hash collisions,address was installed
         if (address_tab[index] == address) begin
            readdata = data_tab[index];
            disable search_address;
         end
         else if (address_tab[index] === 'hxxxxxxxx) begin
            $display (
   "%m: Trying to read uninitialized data at address %h",
               address);
            $stop;
            disable search_address;
         end
      end
      $display ("%m: Hash table is full");
      $stop;
   end
end

endfunction
```

**Figure 9.14** Accessing a memory location using the hash method

show the tasks for accessing and modifying memory locations using the hashing technique.

The `readdata` function handles hash collisions in the `search_address` block. When two addresses hash to the same location, a simple (though not necessarily efficient) method is used to resolve the conflict: a linear scan of the table to find a free location. If none is available, the hash table is full and a message is printed. The model stops at this point.

```
task writedata;
input [31:0] address;
input [31:0] value;
integer index, i;

begin
   index = hash (address);
   if (address_tab[index] == address)
      data_tab[index] = value;
   else begin : search_address
      // The first hashing did not produce the right address
      // Start scanning the table for the address
      for (i = (index + 1) % 'HASHSIZE;
           i != index;
           i = (i + 1) % 'HASHSIZE) begin
         if (address_tab[index] == address) begin
            //Due to hash collisions, address was installed
            disable search_address;
         end
         else if (address_tab[index] === 'hXXXXXXXX) begin
            data_tab[index] = value;
            address_tab[index] = address;
            disable search_address;
         end
      end
      // The whole table was searched for an empty slot
      $display ("%m: Hash table is full");
      $stop;
   end
end

endtask
```

**Figure 9.15** Modifying a memory location using the hash method

## Loading Interleaved Memory

Normally, when modeling a memory subsystem, the contents of memory reside in a disk object file and are loaded into a Verilog memory array during initialization. The compiler generates the object code assuming a single sequential and contiguous memory. In that case, it can be loaded into internal arrays using Verilog's $readmemb or $readmemh task. However, simulating interleaved memory requires some special steps.

Let us consider a two-way interleaved memory. The concept is easily extended to multiway interleaving. A 256-word memory, mem,

## Useful Modeling and Debugging Techniques

can be implemented by two 128-word memories, mem1 and mem2, where mem1 holds even addresses (0,2,4,...126) and mem2 holds odd

```
module interleaved_memory;
parameter       DSIZE = 32,
                ASIZE = 8,
                MAXWORDS = 1 << ASIZE,   // = 256
                HALFMAX = MAXWORDS/2 ;

reg [DSIZE-1:0] MEM[0:MAXWORDS-1],
                MEM1[0:HALFMAX-1],
                MEM2[0:HALFMAX-1] ;

integer i, j;

task memaccess (address, data, rw) ;
input [ASIZE-1:0] address ;
inout [DSIZE-1:0] data ;
input rw ;
reg   [ASIZE-2:0] addr ;

begin
   addr = address[ASIZE-1:1] ;
   if (address[0] == 0) begin : even_address
      if (rw == 1) // even_read
         data = MEM1[addr] ;
      else         // even_write
         MEM1[addr] = data ;
   end
   else begin : odd_address
      if (rw == 1) // odd_read
         data = MEM2[addr] ;
      else         // odd_write
         MEM2[addr] = data ;
   end
end

endtask

   initial begin : loadmem
      j = 0 ;
      $readmemh("PROG.FILE",MEM) ;
      for (i=0; i < MAXWORDS; i=i+2) begin
         MEM1[j] = MEM[i] ;
         MEM2[j] = MEM[i+1] ;
         j = j + 1 ;
      end
   end
endmodule
```

**Figure 9.16** A model for loading interleaved memory

addresses (1,3,5,...127). The size of each word is identical in mem, mem1, and mem2. This is called "horizontal interleaving".

The model shown in Figure 9.16 takes the object code from a file prog.file and loads it into MEM1 and MEM2 with even and odd addresses respectively. The memaccess task reads and writes data to or from the appropriate memory bank, given an address and the type of the operation (read or write). The least significant address bit, address[0], is decoded to determine in which memory bank the address will be found. This procedure is equivalent to asserting a chip select signal for one of the two memory banks from the least significant bit and using the rest of the high order bits, address[ASIZE-1:1], to access the appropriate location.

The memaccess task does not have to be as trivial as shown here. A complete handshake protocol for memory access can be implemented as a part of this task. Nevertheless, no matter how complex the task becomes, it provides a transparent interface between the memory implementation and the rest of the system.

The description shown in Figure 9.16 is inefficient. This becomes evident when the memory size is very large. Notice that we declared three memories — MEM, MEM1, and MEM2. The total number of memory words used during the simulation is

```
maxwords+halfmax+halfmax = 2*maxwords
```

or twice the size of a single contiguous memory. A large memory leads to swapping- and paging-related problems because the simulator needs to allocate physical (or virtual) memory to hold the contents of these simulated memories. Unfortunately, in a Verilog simulator, there is no way to "release" a memory. This implies that although MEM was used only for the purpose of loading the contents of the object code from the file into interleaved memory banks, it occupies runtime memory of the simulator but remains unused.

To work around the simulator problem, one approach is to write a simple C program that generates two files, prog.file1 and prog.file2, by writing out alternate lines of object code from prog.file. Now the declaration for MEM can be deleted and the $readmemh statement can be replaced by the following:

```
$readmemh("PROG.FILE1", MEM1) ;
$readmemh("PROG.FILE2", MEM2) ;
```

# Useful Modeling and Debugging Techniques

```
module splitfiles ;
parameter DSIZE = 32 ,
          ASIZE = 8,
          MAXWORDS = 1 << ASIZE ;
integer i, f1, f2 ;

   reg [DSIZE-1:0] MEM[MAXWORDS-1:0] ;

   initial begin
       $fopen("PROG.FILE1",f1) ;
       $fopen("PROG.FILE2",f2) ;
       $readmemh("PROG.FILE",MEM) ;
       for (i=0; i<MAXWORDS; i = i+2) begin
           $fdisplayh(f1,MEM[i] ;
           $fdisplayh(f2,MEM[i+1] ;
       end
end
endmodule
```

**Figure 9.17** An efficient model for loading interleaved memory

The example shown in Figure 9.17 accomplishes the same task in Verilog instead of in C. In either approach it is a two-step process: first convert a single object code file into two object code files, and then read these object code files into the appropriate memory banks at the start of the simulation. This approach eliminates the use of mem completely, using only the required simulated and physical memories.

## Verification of Setup and Hold Constraints

In many design projects, you have to write not only the module being designed but also the external environment around the module. For example, when designing a microprocessor chip, you may have to design a model for the memory subsystem, the disk subsystem, the keyboard, or the console. The purpose of this environment is to generate stimuli for the module being designed and to collect and display simulation results.

Many times the environment models must check the timing relationships among signals on the I/O ports of the module being designed. For example, if the module is a Micro Channel bus controller, then it has to obey some setup and hold constraints. Such checks can be isolated in separate processes (always loops) which continuously monitor the signals in question and their relationships. Chapter 8 shows the use of such checks where the floppy disk controller, acting as part of

```
module m;
   reg a, b;
   wire c = a & b;
   initial begin
      a = 1;
      b = 1;
      //#0 ;   Enable this statement to see c modified
      $display ("a = %b, b = %b, c = %b", a, b, c);
   end
endmodule
```

**Figure 9.18** Effect of Verilog event scheduling

the simulation environment, checks the correctness of the signals produced by the host CPU interface.

## Effects of Verilog Execution Order and Scheduling

A Verilog model has to be analyzed in the context of the Verilog scheduling algorithm. The real hardware operates in real parallelism; however, simulating the hardware on a machine with a single CPU implies that two events cannot occur simultaneously, even if they occur at the same simulated time. This phenomenon can be confusing, and the following two examples demonstrate its effect. Consider the Verilog model in Figure 9.18.

Although the expression a & b is continuously assigned to wire c, the assignment does not take effect immediately when a or b changes. Instead, the Verilog simulator just schedules an event that recomputes the value of c whenever a or b changes, and continues to execute the next statements until it encounters a timing control construct (@, #, or wait). At that time the simulator executes the event to compute the value of c.

In order to see the effect of the continuous assignment, simply insert a zero delay statement (#0 ;) just before the $display statement. This reschedules the $display statement to the end of the current simulation time, by which time the effect of changing a and b has propagated to c.

## Useful Modeling and Debugging Techniques

```
module m;
integer a, b, c;
event    e;

   always @e begin : e_block
      c = a + b;
   end

   initial begin
      a = 1;
      b = 2;
      ->e;
      // #0 ;// Enable this statement to see c modified
      $display ("a = %d, b = %d, c = %d", a, b, c);
   end

endmodule
```

**Figure 9.19** Effect of Verilog event scheduling

Figure 9.19 illustrates another example of the same phenomenon. Here we trigger an event that is supposed to modify a variable. Triggering the event just schedules the execution of the task but does not yet execute the task. Here too, inserting a zero delay statement immediately before the `$display` statement executes the e_block.

```
module comb (in1, in2, in3, in4, in5, out1, out2);
input   in1, in2, in3, in4, in5;
output  out1, out2;
        assign { out1, out2 } = in1+in2+in3+in4+in5;
endmodule

module top;
reg   [4:0]  inreg;
wire  [1:0]  outwire;
integer      i;

parameter  testsize = 5000;

   comb c (inreg[4], inreg[3], inreg[2], inreg[1], inreg[0],
           outwire[1], outwire[0]);
   initial
      for (i = 0; i < testsize; i = i + 1)
         inreg = i;
   always @outwire
      $display ("outwire = %b", outwire);

endmodule
```

**Figure 9.20** Test for combinational module execution speed

327

Another common situation where event scheduling can be misleading is where a module has both behavioral and structural instances. Suppose that you have a combinational block that you want to test for speed. One way to do this is to write a top module that instantiates one copy of the module under test, and to have a behavioral instance that generates the input test patterns as shown in Figure 9.20.

If you run the module as shown in Figure 9.20 you will discover that it executes very quickly, and that the $display statement is executed only once at the end of the simulation. The loop that assigns the input to the combinational module executes as a single event, so the combin-ational block never gets control to execute its function until the loop terminates and relinquishes control to the event scheduler. By modifying the statement

```
inreg = i;
```

to

```
#0 inreg = i;
```

you can remedy the problem, and the test will execute as intended.

```
task readmem;
input   [MEMSIZE:0] addr;
output [DATSIZE:0] val;
begin
   abus = addr;
   #1 cs_ = 0;
   #1 rd_ = 0;
   #1 val = dbus;
   rd_ = 1;
   cs_ = 1;
   #1 ;
end
endtask

task writemem;
input   [MEMSIZE:0] addr;
input [DATSIZE:0] val;
begin
   abus = addr;
   cs_ = 0;
   dbusreg = val;
   #1 wr_ = 0;
   #1 wr_ = 1;
   #1 cs_ = 1;
   dbusreg = 8'hzz;
end
endtask
```

**Figure 9.21** Basic test tasks for memory testing

## Useful Modeling and Debugging Techniques

```
task initmem;
input start, finish, pattern;
integer start, finish, pattern;
begin : initmem_block
integer i;
    for (i = start; i <= finish; i = i + 1)
        writemem (i, pattern+i);
end
endtask

task copymem;
input source, dest, size;
integer source, dest, size;
begin : copymem_block
integer i;
reg[DATSIZE:0] tmp;
    for (i = 0; i <= size; i = i + 1) begin
        readmem (source + i, tmp);
        writemem (dest + i, tmp);
    end
end
endtask

task comparemem;
input source, dest, size;
integer source, dest, size;
begin : comparemem_block
integer i;
reg[DATSIZE:0] tmp1, tmp2;
    for (i = 0; i <= size; i = i + 1) begin
        readmem (source + i, tmp1);
        readmem (dest + i, tmp2);
        if (tmp1 != tmp2) begin
            $display
                ("%m: error,mem[%h](%h)!=mem[%h](%h)",
                source+i, tmp1, dest+i, tmp2);
            $stop;
        end
    end
end
endtask
```

**Figure 9.22** Second-level test tasks for memory testing

## Generation of Test Vectors for Complex Modules

The best way to obtain test data is from a real system. For example, in a model of a computer system, you can load the memory with a test program and begin executing the program on the model. Sometimes, you may get a trace file from running the actual device in a test environment while collecting the data (e.g., with a logic analyzer). More often than not, you must write your own test program. It is usually advantageous to develop the tests from the bottom up, writing small tasks to perform trivial operations and stringing them into more complicated tasks.

We exemplify this technique by developing a test for a memory subsystem module. The most basic operations on the memory are read from an address and write to an address. Figure 9.21 shows the code to implement the two tasks. It assumes the control signals cs_ (chip select), rd_ (read), and wr_ (write).

```
task test;
input start, finish, pattern;
integer start, finish, pattern;
begin : test_block
integer size, middle;
   size = (finish - start) / 2;
   middle = start + size;
   initmem (start, finish, pattern);
   copymem (start, middle, size);
   comparemem (start, middle, size);
end
endtask
```

**Figure 9.23** Driver for memory testing

Having written the basic tasks, we can easily write more complicated tasks that are based on them. For our purpose we need three compound tasks: one for initializing the memory, one for copying a block from one location in memory to another, and one for comparing two blocks of memory. All of these tasks are shown in Figure 9.22.

Finally we use the second-level tasks to build even a higher level task. This task can perform a complete test by initializing the memory, copying a block of memory to a different address, and comparing the source with the destination (Figure 9.23).

# Useful Modeling and Debugging Techniques

```
module test_mem;
parameter
   MEMSIZE = 10,
   DATSIZE = 8,
   MAXADDR = (1 << MEMSIZE) - 1;
wire [DATSIZE-1:0] dbus;
reg [MEMSIZE-1:0] abus;
reg [DATSIZE-1:0] dbusreg;
reg cs_, wr_, rd_;

mem m (rd_, wr_, cs_, abus, dbus);

task readmem;
...
endtask

task writemem;
...
endtask

...

initial begin
   test (0, 100, 0);
   test ('hff, 'hfff, 'haa);
   ...
end

endmodule
```

**Figure 9.24** Skeleton of the full test module

A skeleton for the full module is given in Figure 9.24. Increasing the nesting depth of task calls could affect the efficiency of simulation. However, it is usually preferable to defer efficiency considerations to a later stage of simulation, when speed becomes more important. It is much easier to take a readable model and improve its efficiency than to start with a poorly written model and make it more readable and maintainable.

## Verification of the Test Vectors

In many cases, test vector generation is only part of the test generation problem. Analyzing test results and verifying their correctness is also a major task. If you have a way to generate the expected results, you can use Verilog or some utility to automatically compare the expected results with the actual ones. If you do not have an easy way to generate the expected results, then, by judiciously selecting the test vectors, you can simplify the process of inspecting and verifying the simulation results. Of course, test vector generation is case specific; but we can demonstrate the technique by using an example.

Consider a barrel shifter that is a part of an ALU:

```
module shifter (
        in32,    // 32 bit input
        cin,     // carry in
        shft,    // 5 bit shift amount
        byte,    // byte operation (8 bit)
        word,    // word operation (16 bit)
        double,  // double word operation (32 bit)
        dir,     // direction (0 - left, 1 - right)
        op,      // 2 bit operation:
                 //   00 - shift logical
                 //   01 - shift arithmetic
                 //   10 - rotate
                 //   11 - rotate with cary
        out32,   // 32 bit output
        cout     // carry out
);
```

Clearly the size of the input prohibits exhaustive testing. Using random tests makes it excruciatingly difficult to confirm the correctness of the results. Instead we will design the test based on the expected result.

The shifter has four operations which have to be tested in two directions, with two different carry bit values, and with three different input widths (byte, word, and double word mode)—a total of 4*2*2*3 = 48 sections. A typical output section for, say, left shift arithmetic with carry set and byte mode looks like the example shown in Figure 9.25.

```
in32 = 00000000000000000000000010000000
cin = 1, b = 1, w = 0, d = 0
dir = 1 (left), op = 01 (shift arithmetic)

shift   cout                out32
----    ----                -----
  0      1       00000000000000000000000010000000
  1      1       00000000000000000000000011000000
  2      1       00000000000000000000000011100000
  3      1       00000000000000000000000011110000
  4      1       00000000000000000000000011111000
  5      1       00000000000000000000000011111100
  6      1       00000000000000000000000011111110
  7      1       00000000000000000000000011111111
  8      1       00000000000000000000000011111111
```

**Figure 9.25** Test vectors organized for ease of inspection

# Useful Modeling and Debugging Techniques

The pattern in the figure is regular and can be easily verified by visual inspection.

The other sections of the test can be designed in a similar way, and the printout can be set in a way that shows at a glance that the output is correct.

## Summary

This chapter presented a collection of techniques that can be useful when writing Verilog models. As you gain experience, you will be able to develop your own techniques. But, while writing more and more complex models, don't forget that the most important quality of good Verilog code is its readability. Use meaningful variable names, modularize your code into tasks and functions, use indentation to show the block structure, and use comments to explain the functionality of your code.

# APPENDIX A

# Condensed Language Reference Manual

## Syntax Conventions

We use a special convention - extended BNF, or EBNF - to describe the syntax of the language. Each production rule in the syntax has the form

```
<left hand side symbol>
    -> alternative
    -> alternative ...
```

The following symbols are reserved and have special meaning in EBNF:

|       Or (and alternate choice).
[ ]    Optional symbols.
( )    Group of symbols
...    Repeatable symbol (one or more occurrences).
\?...] Repeatable symbol with separator (one or more occurences).
[ ]... Optional repeatable symbol (zero or more occurrences).
( )... Repeatable group (one or more occurrences).
[ ]?...Optional repeatable symbol with separator (zero or more occurrences).

335

For example, the rule

```
<module header>
        -> '(' [<port>] ')' ';'
        -> ';'
```

is equivalent to

```
<module header>
        -> '(' ')' ';'
<module header>
        -> '(' <port> ')' ';'
<module header>
        -> '(' <port> ',' <port> ')' ';'
<module header>
        -> '(' <port> ',' <port> ',' <port> ')' ';'
<module header>
        -> '(' <port> ',' <port> <',' port>
        ')' ';'
        ... ...
<module header>
        -> ';'
```

## Lexical Constructs

Lexically, Verilog is very similar to C or C++. The language is case-sensitive and all its keywords are lower case. White space - spaces, tabs and new-lines is ignored. Verilog has two types of comments: one-line comments start with "//" and end at the end of the line. Multiline comments start with "/*" and end with "*/".

Variable names must begin with an alphabetic character or underscore followed by alphanumeric or underscore characters. The exception is that system tasks and functions start with a dollar sign. The Verilog name can include non alphanumeric characters if the first character is a backslash (\). In that case the name includes all the characters following the backslash until the first white space character; namely, until a space, tab or new line character.

# Condensed Language Reference Manual

Integer literals can have underscores embedded in them for improved readability. The underscores are otherwise ignored. Integer literals can have one of the following formats:

```
<integer_literal>
    -> <integer>
    -> <base> <integer>
    -> <width> <base> <integer>
```

where `<width>` is a base 10 integer specifying how many bits in the literal, `<base>` denotes the representation of the literal as follows: `'b` or `'B` means binary representation, `'o` or `'O` means octal representation, `'d` or `'D` means decimal representation and `'h` or `'H` means hexadecimal representation.

A binary integer literal can be composed of the characters "`01zZxX`", an octal literal can be composed of "`01234567zZxX`" a decimal literal can be composed of "`0123456789`", and a hexadecimal literal can be composed of "`0123456789abcdefABCDEFzZxX`".

The values z and Z stand for high impedance, and x and X stand for uninitialized variables or nets with conflicting drivers. If `<base>` is omitted, then decimal base is assumed. If `<width>` is omitted, then the width of the literal is implied by the context in which it appears.

String symbols are enclosed within double quotes ( " ), and can not span multiple lines. Real number literals can be either in fixed notation or in scientific notation, e.g. `0.5` or `5.0e-1`.

Verilog has compiler directives that affect the processing of the input files. The directives start with a grave accent ( ' ) followed by some keyword. A directive takes effect from the point that it appears in the file until either the end of all the files or until another directive that cancels the effect of the first one is encountered. The effect of a compiler directive can span multiple files. Verilog has many compiler directives, but here we shall only cover `'define`, `'include` and `'ifdef`.

```
<compiler_directive_define>
    -> 'define <macro_name> <replacement_text>
```

This defines a macro called `macro_name`. When the text `'macro_name` appears in the text, it is replaced by the replacement_text. If the macro definition line has the line comment characters in it (//), the comment is discarded and is not part of the replacement_text. Verilog macros are simple text substitutions and do not have arguments (unlike C macros).

```
<compiler_directive if>
    -> 'ifdef <name> <Verilog_code> 'endif
```

If "name" is a defined macro, then the Verilog code until `'endif` is inserted for the next processing phase. If "name" is not a defined macro, the code is discarded.

```
<compiler_directive_include>
    -> 'include <Verilog_file>
```

The code in Verilog_file is inserted for the next processing phase.

## Syntax and Semantics

```
<source_file>
    -> (<module>|<primitive>) *
```

The source file is composed of modules and primitives. Actually, the source also has compiler directives, but these are expanded in a preprocessing pass. The order in which the modules or primitives appear in the file is immaterial, and a module or a primitive can be referenced before it is declared. All the files are processed as if they were concatenated into a single file. Macros take effect from the time that they appear until the end of all the files, and have to be declared before they are referenced.

```
<module>
    -> module <module_name> <port_list>? ';'
        <module_item>*
        endmodule
```

# Condensed Language Reference Manual

A module starts with the keyword " `module` " and ends with the keyword " `endmodule` ". The module_name is unique, and no other module or primitive can have the same name (but the name can be used, however, in other name spaces such as variables, instance names, etc.). The `port_list` is optional. A module without a `port_list` or with an empty `port_list` is typically a top level module, and is used either as the top of the design or for debugging purposes or for testing and observation. Note that there can be more than one top level module in the design.

**User Defined Primitive**

```
<primitive>
    -> primitive <prim_name> <port_list>? ';'
            <primitive_decl>... <primitive_init>?
            table <table_entry>... endtable
            endprimitive
<primitive decl>
    -> input <inp_name>...
    -> output <outp_name>
    -> reg <outp_name>
<primitive init>
    -> initial <outp_name> '=' <init_value> ';'
<table entry>
    -> <combinational_entry>
    -> <lvl_sensitive_sequential_entry>
    -> <edge_sensitive_sequential_entry>
```

A user-defined primitive (UDP) is a special type of a module. Externally a UDP is referenced (i.e. instantiated) exactly like a Verilog gate, such as an AND gate. Internally, UDPs are defined using tables, and therefore they are amenable to optimization. Depending on their implementation, UDPs may simulate faster than their functionally equivalent modules.

UDPs share some properties with Verilog gates. Their ports are one-bit wide, they have exactly one output (always the first port) and one or more input ports. UDPs can be either combinational or sequential. Sequential UDPs are those in which the output is declared as "`reg`", while in combinational UDPs the output is a "`wire`". Sequential UDPs have an internal state, which can be initialized using the "`initial`" declaration. Sequential UDPs can be further divided into level-sensitive

UDPs and edge-sensitive UDPs. In combinational UDPs, the output is a combinational function of the inputs. In a level-sensitive sequential UDP, the output, which is the same as the internal state, is a function of the inputs and of the current internal state. In edge-sensitive sequential UDPs, the output is a function of the inputs, the current internal state, and a transition on one of the inputs.

```
<combinational entry>
     -> <lvl_sym> ':' <outp_sym>
<lvl_sensitive_sequential_entry>
     -> <lvl_sym> ':' <state> ':' <outp_sym>
<state>
     -> <lvl_sym>
<edge_sensitive_sequential_entry>
     -> <edge_inp_list> ':' <state> ':' <outp_sym>
<edge_inp_list>
     -> [<lvl_sym>] <edgc_sym> [<lvl_sym>]
```

The functionality of the UDP is defined by the state table. Each entry in the table defines the output and state of the UDP as a function of the inputs, input transitions, and the previous state.

A `lvl_sym` can be one of the following: "0", "1", "x", "X", "?", "b", or "B". The symbol "?" is a short hand notation for "0", " 1 ", or "x", and the symbols "b" and "B" are short hand notation for "0" or "1".

An `outp_sym` can be one of the following: "0", "1", "x", "?", or "-". The "-" symbol denotes no change. It is only meaningful in sequential UDPs and indicates that the next state and output will remain unchanged.

An `edge_sym` denotes a transition and can be one of the following: (ab) where a and b are level " r ", " f ", " p ", " n ", or " * ". " r " denotes a rising edge and is equivalent to (01) . " f " denotes a falling edge, (10) . " p " denotes a rising edge including unknown and is a short notation for (01) , (0x) or (x1) . " n " denotes a negative edge including unknown, namely (10), (1x) or (x0) . " * " denotes all transitions and is equivalent to (??) . Note that there can be only one edge transition column in a UDP table. That is, an edge-sensitive UDP can be sensitive to an edge on only one of its inputs.

The table below summarizes all the possible table entry symbols.

| Notation | Representation |
|---|---|
| 0 | logic 0 |
| 1 | logic 1 |
| x | unknown |
| ? | either 0, 1, x (input ports only) |
| b | either 0 or 1 (input ports only) |
| - | no change (outputs of sequential primitives) |
| (xy) | value change x, y = 0, 1, x, ?, or b |
| * | any value change (same as (??)) |
| r | rising edge on input (01) |
| f | falling edge on input (10) |
| p | postive edge ((01), (0x) or (x1)) |
| n | negative edge ((10), (1x), (x0)) |

```
<port_list>
        -> '(' <port> ')'
```

The `port_list` is a (possibly empty) list of ports separated by commas. Each port has a width and each net connected to a port has a type (input, output or inout). Ports of primitives are one bit wide; the first port of a primitive is an output and the rest are inputs. Modules have no such restriction.

```
<port>
      -> <port_name>
      -> <internal_bus> <rangeorbit>
      -> '.' <port_name> '(' <lvalue> ')'
<port name>
      -> <identifier> <Internal_bus>
      -> <identifier> <lvalue>
      -> <variable> -> <bit_select>
      -> <part_select>
      -> <lvalue_concatenation>
<lvalue_concatenation>
      -> ' ' <lvalue> ' '
```

An *lvalue* is an expression that can be on the left hand side (LHS) of an assignment, the LHS of a continuous assignment, or as a port expression in an input port. It can consist of a variable name, a single bit selected from a variable, a range selected from a variable or a concatenation of any of the above.

A port can have three forms: a regular port, an unnamed port or a renamed port. In the great majority of cases, a port is simply an identifier which designates the name of the port. An unnamed port is an identifier followed by a range or a bit selection, e.g. `data_bus[11:9]` or `address[8]`. In this case the port itself does not have a name, but is connected to some internal bus. When such a module is instantiated, the port connections can be done by position only and not by name. A renamed port is a dot followed by the port name, followed by an lvalue which represents some internal signals, e.g. `.shortaddress (address[7:0])`. In this case the port is connected directly to these internal signals. The port name is possibly different from the name of the internal signal to which it is connected. Note that a port can be connected to several internal signals, some of them inputs, some of them outputs and some of them inouts. In other words, the direction (input, output or inout) is not a property of the port, but rather the property of the internal signals to which the port is connected.

## Condensed Language Reference Manual

```
<module_item>
        -> <param_decl>
        -> <inp_decl>
        -> <outp_decl>
        -> <inout_decl>
        -> <net_decl>
        -> <reg_decl>
        -> <integer_decl>
        -> <real_decl>
        -> <time_decl>
        -> <event_decl>
        -> <gate_instantiation>
        -> <module_instantiation>
        -> <primitive_instantiation>
        -> <continuous_assign>
        -> <always_block>
        -> <initial_block>
        -> <function_decl>
        -> <task_decl>
        -> <specify_block>
```

Module items can appear in any order inside the module. However, most elements can be referenced only after they have been declared. The exceptions are functions and tasks which can be referenced before they have been declared.

```
<param decl>
        -> parameter <range>? <param assgn> ;
<param assgn>
        -> <param name> '=' <const expr>
```

Each parameter declaration defines named integer constants and gives them initial default values. These default values can be overridden when the module is instantiated, such that each instance may have different parameter value than the other instances. Parameters can appear in any place in the module where an integer constant can appear, for example as a delay value or as the width of a register. If a range is missing then the parameter will have an implementation dependent default width (typically 32 bit). The default value of the parameters can be overridden on a per instance basis using the `defparam` construct (to be discussed later). The `const` in the `param` can be any compile time expression and can contain other parameters.

```
<inp_decl>
     -> input <range>? <var_list> ';'
<outp_decl>
     -> output <range>? <var_list> ';'
<inout_decl>
     -> inout <range>? <var_list> ';'
<var_list>
     -> <variable>
```

`Input`, `output`, and `inout` declarations give the width and type of the module port nets. Each port net must be in exactly one such declaration. If the range is omitted, then it is defaulted to one bit. If a range exists, then it applies to all the variables in the declaration. `var_list` is a list of variables separated by commas.

```
<net_decl>
     -> <net_type> <chrg_strng>? <range>?
              <delay>? <var_list> ';'
     -> <net type> <drive_strng>? <range>?
              <delay>? <assign_list> ';'
<net_type>
     -> wire
     -> wand
     -> wor
     -> tri
     -> triand
     -> trior
     -> trireg
     -> tri0 -> tri1
     -> supply0
     -> supply1
<chrg_strng>
     -> '(' small ')'
     -> '(' medium ')'
     -> '(' large ')'
<delay>
     -> '#' <const_expr>
     -> '#' '(' const_expression ')'
```

`net_decl` declares a group of nets. Nets correspond to physical wires that connect instances. The default range of a wire is one bit. Nets do not store values and have to be continuously driven. If a net has multiple drivers (for example two gate outputs are tied together), then the net value is resolved according to its type. A wand net or triand net operates as a wired AND, and a wor net or trior net operates as a wired OR. A wire or a tri net has a more complex resolution function: If all the drivers have the same value then the wire resolves to this value. If all the drivers except one have a value of `'z` then the wire resolves to the non `'z` value. If two or more non `'z` drivers have different drive strength, then the wire resolves to the stronger driver. If two drivers of equal strength have different values, then the wire resolves to `'x`. A trireg net behaves like a wire except that when all the drivers of the net are in high impedance (z) state, then the net retains its last driven value. `triregs` are used to model capacitive networks. `tri0` and `tri1` nets model nets with resistive pulldown or pullup devices on them. When a `tri0` net is not driven, then its value is 0. When a `tri1` net is not driven, then its value is 1. `supply0` and `supply1` model nets that are connected to the ground or power supply.

Charge strength is associated with capacitive nets and provides small grain resolution when these nets are tied. `delay` is a constant which specifies the propagation delay of the signal in the net. `delay` can consist of one, two or three expressions. If one expression is specified then this is the delay for any transition of the net value. If two expressions are given, then the first one applies to the rise time delay and the second applies to the fall time delay. If three expressions are given, then the first two apply to the rise and fall delays respectively, and the third one applies to the turn off (z) delay.

Single bit nets can be declared implicitly by simply using their names in expressions. In that case, the net type is wire unless a compiler directive has specified a different default type for undeclared nets.

Net declarations can be of two different forms. The first form merely declares the net names. The second form declares the name of the net and also provides a driver for the net. This second form is a shorthand which combines the declaration of the nets with their assignment. Each declaration of the form

```
wire w = a & b;
```

is equivalent to the two constructs

```
wire w; assign w = a & b;
```

The `assign` construct (also called continuous assignment) will be discussed later.

```
<reg_decl>
      -> reg <range>? <reg_var_list> ';'
<reg_var_list>
      -> <reg_variable>
<reg_variable> -> <variable> <range>?
```

`reg_decl` declares a group of `reg` variables with default width of 1. As opposed to `net` declarations, `reg`s can form arrays (this is the only data type in Verilog that can form an array). Although `reg` is a shorthand for register, `reg`s do not automatically imply storage elements. As opposed to `net` variables which need to be continuously driven, `reg` variables get their value through assignments and store the value between assignments. The assignment will typically be performed in behavioral instances (such as `initial` or `always` blocks), in functions or in tasks. Input ports may not be of type `reg`, because they are continuously driven from the outside. For the same reason, `reg`s cannot be on the left hand side (LHS) of a continuous assignment (to be discussed later), nor can they be connected to the output of a child submodule. Further discussion of the difference between `reg`s and `net`s will be given later. `Reg`s and `net`s can appear in expressions wherever integers can appear, but `reg`s and `net`s are always treated as unsigned two's-complement entities as opposed to integer variables which are treated as signed two's complement variables. For example the expression $r < 0$ will always evaluate to false if $r$ is a `reg` or a `net`.

```
<integer_decl>
      -> integer <var_list> ';'
<real_decl>
      -> real <var_list> ';'
<time_decl>
      -> time <var_list> ';'
<event_decl>
          -> event <var_list> ';'
```

These declare groups of integers, reals, time variables and events respectively. Integers are machine dependent sized integers (typically 32 bits). Reals are also machine dependent and represent floating point numbers. Time variables are unsigned integers, which are also machine dependent but typically are 64-bit wide. They are used to hold and manipulate simulation time values. Events are special variables that hold event names. Such events can be triggered inside behavioral instances and can synchronize the operation of two parallel executing blocks.

Each variable in a module such as net, reg, integer, real, time and event has to have a unique name in its scope. A scope is typically a module, but scopes can be nested when a named block is declared. This will be discussed later.

```
<gate_instantiation>
    -> <gate_type> <drive_strng>? <delay>?
         <gate_instance> ';'
<gate_instance>
    -> <instance_name>? '('
         <terminals> ')' <drive_strng>
    -> '(' <zero_strng> ',' <one_strng> ')'
    -> '(' <one_strng> ',' <zero_strng> ')'
```

Gate instantiation starts with a gate type, followed by optional drive strength, followed by optional delay, followed by a list of gate instances separated by commas. The gate type is one of the pre-defined Verilog primitives: and, nand, or, nor, xor, xnor, buf, not bufif0, bufif1, notif0, notif1, nmos, pmos, cmos, rnmos, rpmos, rcmos, tran, tranif0, tranif1, ntran, ntranif0, ntranif1, pullup, pulldown. Individual instances can be named, or they can be anonymous. Each instance has a list of ports surrounded by parentheses. In all Verilog primitives and in user defined primitives, the output is the first port, and the inputs follow. Also, all primitive and UDP ports are single bit wide. Some of the Verilog primitives, AND, NAND, OR, NOR, XOR and XNOR gates, have variable number of inputs.

Drive is used to resolve the final value of a net which is driven by more than one driver. Zero can be one of supply0, strong0, pull0, weak0, and highz0, where supply0 is the strongest and highz0 is the weakest. Similarly, one can be one of `supply1`, `strong1`, `pull1`, `weak1`, and `highz1`. The default value for drive is (`strong0`, `strong1`).

```
<module_instantiation>
    -> <module_name> [<param_values>]
         <module_instance> ';'
<param_values>
    -> ' ' '(' <const_expr> ')' <module_instance>
    -> <instance_name> '(' <terminals>? ')'
<primitive_instantiation> -> <prim name>
         <drive_strng>? <delay>?
         <primitive_instance> ';'
<primitive_instance>
    -> <instance_name>? '('
           <terminals>? ')'
```

Module and user defined primitive (UDP) instantiations start with the module or UDP name followed by a list of instances separated by commas. Each instance starts with an instance name followed by a list of terminals surrounded by parenthesis. The module or UDP name has to correspond to one of the module or UDP definitions. The corresponding module definition can come before or after the instance, but recursion, direct or indirect, is not allowed. The instance names have to be unique in the current module, i.e. no two instances of modules, UDP or Verilog primitives in the same module can have the same name. Note that Verilog primitive and UDP instances can be anonymous, but module instances need to be named.

If parameter_values exist in the module instantiation, then the expressions in the list are evaluated and they replace the values of the parameters in the module definition. The replacement is done by position, and parameters cannot be skipped. Note that the syntax for gate and UDP instantiation is very similar if the instance delays are considered as parameters.

```
<terminals>
    -> <positional terminal>
    -> <named terminal>
<positional terminal>
    -> <expr> -> <blank> <named terminal>
    -> '.' <port name> '(' <expr> ')'
```

## Condensed Language Reference Manual

Terminals in instance constructs designate the connection between the module in which the instance is placed and the port of the instance. Positional terminals connect the instance terminals by position, as defined in the module declaration for the instance. The first port is connected to the first expression, the second port to the second expression and so on. Blanks can be used to leave some ports unconnected. Named terminals connect expressions to instance ports by name. If the port is an output or inout port, then the expression has to be an *lvalue*, but input ports can be arbitrarily complex expressions. In any case the expression has to be of the same width as the port width.

```
<continuous_assign>
        -> assign <drive_strng>? <delay>?
                <assignment> ';'
<assignment>
        -> <lvalue> '=' <expr>
```

Assign declares a sequence of "continuous assignment" constructs. Each continuous assignment continuously monitors the variables in the expression on the right hand side (RHS), and when any of them changes the expression is evaluated and the result is assigned to the LHS. Continuous assignment is one way of describing combinational logic, and is often referred as the "data-flow" style of modeling. The expression can be arbitrarily complex and can involve function calls. If more than one variable changes at the same time, the expression may be evaluated more than once. This is implementation dependent and there is a potential for optimization. The simulation result should not be affected by such multiple evaluations. Because the LHS of the continuous assignment is continuously driven, it has to consist of net variables and not reg variables.

If the LHS expression of the continuous assignment includes function calls, then the expression will be re-evaluated whenever any of the arguments of the function changes. However, a change in other

variables that are used in the function but are not in the argument list will
not cause the re-evaluation of the expression.

```
<expr>
        -> <operand>
        -> <unary_operator> <expr>
        -> <expr> <binary_operator> <expr>
        -> <bool_expr> '?' <expr> ':'
               <expr> <operand>
        -> <number> -> <identifier>
        -> <identifier> '[' <expr> ']'
        -> <identifier> '[' <const_expr> ':'
              <const_expr> ']'
        -> <concat> -> <duplication>
        -> <function_call>
        -> '(' <expr> ')' <concat>
        -> ' ' <expr> ' ' <duplication>
        -> ' ' <const_expr> ' ' <concat> ' ' ' '
<function_call>
        -> <function_name> '(' <expr> ')'
<function_name>
        -> <identifier> verbatim
```

In Verilog, expressions like those in many other programming languages, are used to calculate values. They consist of operands and operators which are applied to the operands using some precedence rules. Parentheses can be used to override the normal precedence rules. The operands in an expression can be any legal `<lvalue>` a constant or a function call. The operators for the most part are the same as the ones in the C programming languages, but there are a few exceptions. Verilog does not have the increment/decrement operators (++ and --) and does not have the combined assignment operators (+=, +-, etc.). Verilog has the following unary reduction operators: &, |, ~&, ~|, ^, ~^, and ^~. When one of these operators is applied, it is inserted between all the bits of the operand. For example, ^a will return the parity of a.

The bit select operator, e.g. a[i] has the same syntax whether a is a vector — in which case i selects a single bit from the vector — or whether a is an array of regs, in which case i selects one reg from the array. The range select operator, e.g. a[i:j] selects a subfield from a vector.

## Condensed Language Reference Manual

The concatenate operator, e.g. {a, b, c} concatenates its arguments into a single vector, and can be used either as an operand in an expression, or on the LHS of an assignment. If a literal (integer) is one of the arguments to a concatenation, then it has to have an explicit width.

The duplication operator has two arguments, a positive integer and a concatenation expression. The operator concatenates its right argument to itself as many times as specified by the left argument, e.g. 5{a, b} yields a, b, a, b, a, b, a, b, a, b.

The table below summarizes the binary operators and their precedence.

| Operator | Precedence |
|---|---|
| +, -, !, ~ (unary) | Highest |
| *, / % | |
| +, - (binary) | |
| <<. >> | |
| <, <=, >, >= | |
| =, ==. != | |
| ===, !== | |
| &, ~& | |
| ^, ^~ | |
| \|, ~\| | |
| && | |
| \|\| | |
| ?: | Lowest |

```
<always_block>
     -> always <timing_ctrl>?... <statement>
<initial_block>
     -> initial <timing_ctrl>?... <statement>
```

351

Behavioral instances are of two types: `always` blocks and `initial` blocks. `always` blocks start execution at the beginning of the simulation and continue execution in an infinite loop until the end of simulation. `initial` blocks also start execution at the beginning of the simulation but when execution reaches the end of the block, the thread of execution stops. Because `always` blocks execute in an infinite loop, each such block needs to have at least one timing control construct or one wait statement in it, otherwise the execution of the block will continue forever, in an infinite loop, and simulation time will never advance.

```
<timing ctrl>
     -> '@' <event expr>
     -> ' ' <delay expr> <event expr>
     -> <event>
     -> '(' <event> ')' <event>
     -> <event variable>
     -> <nonevent variable>
     -> '(' posedge <variable> ')'
     -> '(' negedge <variable> ')'
<delay expr>
     -> <expr>
```

Timing control constructs cause the following statement or block execution to be suspended. Timing control constructs can be of two types: "at delays" and "relative delays". An "at delay" causes execution to be suspended until a specific event occurs. A "relative delay" causes execution to be suspended for a specified period of time. An "at delay" can wait for different types of events. If the event expression is a single (possibly multi-bit) non event variable, then execution will resume whenever the variable has changed state. If the event expression is an event variable, then the execution will resume when this event has been triggered (see below). If the event expression has the form of (`posedge` variable) or (`negedge` variable) then variable has to be a single bit net or reg type variable, and the execution will resume if the variable goes through a positive or negative transition respectively. An event expression composed of several events separated by `or` will resume execution when any one of the events triggers. Note that you can not use the bitwise OR operator (" | ") in an event expression, but instead, you need to use the keyword or . Also note that there is no similar and operator for events.

```
<timed statement>
        -> <timing ctrl>?... <statement> verbatim
<function decl>
        -> function <range>? <function name> ';'
              <taskfunc decl>?... <statement>
              endfunction
<task decl>
        -> task <task name> ';' <taskfunc decl>?...
              <statement> endtask
```

A function is used to encapsulate a piece of code that calculates a value. A function takes some input arguments - at least one input is required - and calculates its return value. The return value has a width which defaults to one bit. The function returns a value by assigning the value to the function variable.

Tasks are used to encapsulate pieces of behavioral code. They take zero or more arguments which can be inputs, outputs or inouts. Tasks do not return a value but they can cause side effects by modifying their output arguments. Incidentally, both tasks and functions can cause side effect by modifying variables in the module in which the task or function is defined.

Functions and tasks can declare local variables, and they can use the input arguments, the local variables or global variables in the module to perform their operations. The statements in a function are restricted to those statements that do not have any timing control constructs, including wait statements or task activations, and so functions execute in zero simulation time. For this reason functions can be used to describe combinational elements. Tasks on the other hand can include any statement. Regs that are declared in functions and tasks are generally used as temporary variables and not as memory elements. Note that such regs do hold their values between function or task activations, but usually it is not a good practice to take advantage of this behavior. Functions and tasks are local to the modules in which they are defined, and so if the same function or task is needed in more than one module, it has to be written twice, or it can be invoked using the dot notation (see below).

```
<statement>
        -> <null_statement>
        -> <sequential_block>
        -> <parallel_block>
        -> <procedural_assgn>
        -> <if_statement>
        -> <case_statement>
        -> <repeat_statement>
        -> <forever_statement>
        -> <while_statement>
        -> <for_statement>
        -> <wait_statement>
        -> <trigger_statement>
        -> <task_enable_statement>
        -> <task_disable-statement>
        -> <force_statement>
        -> <release_statement>
        -> <assign_statement>
        -> <deassign_statement>
<null statement>
        -> ';'
```

Statements are the elements of behavioral modeling. They can appear in `always` blocks, `initial` blocks, tasks and functions (but there is some restrictions on which statements can appear in a function). The null statement is needed for example in a task that does nothing except delays its caller. Sequential and parallel blocks are used to group multiple statements into a single statement. This grouping can be nested, and within a block one can declare new variables local to this block. No instantiation can take place inside a block.

```
<sequential_block>
        -> begin <timed_statement>... end
        -> begin ':' <block_name> <block_decl>...
                <timed_statement>... end
<parallel_block>
        -> fork <timed_statement>... join
        -> fork ':' <block_name> <block_decl>...
                <timed_statement>... join
```

Blocks are composed of groups of statements. Sequential blocks execute their statements sequentially and parallel blocks execute their statements in parallel. Multiple threads of executions are created at the point of fork and all the threads execute at the same simulation time. When any of the parallel threads arrives at the join point before the other threads, it waits until all the threads have completed, at which time the threads merge into a single execution thread and then continue. The order in which the threads are executed is implementation dependent.

Blocks can be named or unnamed. If a block is named, then it can have declarations just before the first statement. Each named block creates a new name space, and variables can be declared with names equivalent to those of outer blocks variables. A reference to a variable always refers to the innermost variable with this name. For example, in the code

```
module m;
    always begin: outer
    integer i; i = 1;
        begin: inner
        reg i; i = 0; ...
```

the first assignment to `i` is to the `integer i`, and the second assignment is to the `reg i`.

In all the following constructs, boolean is syntactically equivalent to expression, but semantically, the value of the expression evaluates to TRUE or FALSE as follows: A zero value is FALSE, an x on any of the bits (for a multi-bit expression) evaluates to FALSE, everything else evaluates to TRUE.

The `if`, `for`, and `while` statements are very similar to their equivalent statements in C. The `case` statement is similar to the switch statement in C but each alternative takes only one statement, and there is no fall-through from one alternative to the next.

The `repeat` statement repeats executing its statement a fixed number of times as specified by the value of expression. The forever statement repeats executing its statement forever. A `forever` statement, like an `always` block, needs to have at least one timing control construct, otherwise simulation time will never progress.

```
<if_statement>
      -> if '(' <bool_expr> ')' <statement>
      -> if '(' <bool_expr> ')' <statement>
            else <statement>
<for_statement>
      -> for '(' <assignment> ';' <bool_expr> ';'
            <assignment> ')' <statement>
<case_statement>
      -> <case_type> '(' <expr> ')'
         <case_item>... endcase <case_item>
      -> <expr> ':' <statement>
      -> default ':' <statement>
<repeat_statement>
      -> repeat '(' <expr> ')' <statement>
<forever_statement>
      ->forever <statement>
<while_statement>
      -> while '(' <bool_expr> ')' <statement>
<wait_statement> -> wait '(' <bool_expr> ')'
         <statement>
<trigger_statement>
      -> '->' <event_name> verbatim
```

The `wait` statement is a level sensitive wait, as opposed to the @ construct which is an edge sensitive wait. Execution is suspended until the boolean expression is TRUE before executing the statement. If upon encountering a wait statement the wait condition is true, then execution will continue as a single uninterrupted event.

The trigger statement triggers an event. Once an event has been triggered, all the threads that have been waiting on it, using the @ construct, can now proceed execution.

```
<task_enable_statement>
   -    > <task_name> ';'
        -> <task_name> '(' <expr> ')' ';'
<disable_statement>
        -> disable <task or block_name> ';'
```

A `task_enable` statement starts the execution of the statements in a task after passing the parameters to the task. The next statement after the task will not be executed until the task has completed execution, either normally or through a disable statement. The parameters to the task are matched to the inputs and outputs of the task. The `disable` statement is used to terminate the execution of a task or a named block before it finishes executing its last statement. A task or a block can be disabled from a statement external to it, but many times a disable statement is used internally inside a loop construct to terminate the loop prematurely or to continue with the next iteration of the loop. In this respect, the `disable` statement can function as the break and continue statements in C.

```
<assign_statement>
        -> assign <assignment> ';'
<deassign_statement>
        -> deassign <assignment> ';'
<force_statement>
        -> force <assignment> ';'
<release_statement>
        -> release <lvalue> ';'
```

A procedural assign statement has a similar functionality to that of the continuous assignment construct. This statement connects an *lvalue* to an expression, such that whenever any variable in the expression changes, the expression is re-evaluated and assigned to the `lvalue`. The differences between the continuous assign and the behavioral assign are as follows.

First, syntactically the continuous assignment is instantiated at the module level, whereas a behavioral assign statement can appear only in `always blocks`, `initial blocks` or `tasks`. Second, the constituents of the LHS in the continuous assignments have to be nets, but in behavioral assign statements they have to be regs. Finally the effect of a continuous assignment is permanent throughout the simulation and cannot be changed, in the same way that instances are permanent. Behavioral assign statements can be undone using the `deassign` behavioral statement. If an assign statement is applied to a `reg` for which there is already a procedural assignment, the `reg` is deassigned before the new assign takes effect.

The `force` and `release` statements are similar to the `assign` and deassign statements. They are used to connect an expression to an *lvalue*. But while assign can apply only to `reg` type variables, force can be applied to either `reg` or `net` type variables. Also, `force` is stronger than assign. When a `force` is applied to an assigned `reg`, the `reg` is deassigned and the `force` takes effect. When an `assign` is applied to a forced `reg`, the `assign` is ignored. `Force` also has precedence over continuous assignment, but when a continuously assigned net is released after having been forced, the continuous assignment resumes its effect. `Force` is not usually used for simulation but rather for debugging. A forced `net` or `reg` can be released using the release statement which is analogous to the deassign statement.

```
<procedural_assgn>
        -> <blocking_assgn>
        -> <nonblocking_assgn>
<blocking_assgn>
        -> <lvalue> '=' [<timing_ctrl>] <expr>
<nonblocking_assgn>
        -> <lvalue> '<=' [<repeat_timing_ctrl>] <expr>
<repeat_timing_ctrl>
        -> <timing_ctrl>
        -> repeat '(' <expr> ')' '@' <event_expr>
```

Procedural assignment is the way in which behavioral code can cause side effects. In its simpler form, procedural assignment is very similar to an assignment in other programming languages such as C with the exception that the LHS can be a concatenation of several variables. Note also that the LHS can not be a net type but rather a reg, an integer or a time variable. But in Verilog, the concept of time is essential, and a procedural assignment statement can have one of several flavors, depending on whether this is a blocking or non blocking assignment and depending on whether the optional `timing_ctrl` or `repeat_timing_ctrl` exists. In all cases, the RHS will be evaluated immediately, i.e. at the current simulation time. These cases differ in the way that the assignment is performed and the way in which execution resumes.

A blocking assignment without `timing_ctrl`: The assignment is performed and the new value is available immediately. Execution continues without delay through the next statement and any reference to the assigned variable(s) will yield the new value.

A blocking assignment with `timing_ctrl`: A temporary variable, T, is created and the expression value is stored in T. Execution is delayed until the `timing_ctrl` condition is satisfied, at which time T is assigned to the LHS variables and then execution resumes to the next statement.

A non-blocking assignment without `timing_ctrl`: A temporary variable, T, is created and the expression value is stored in T. Execution continues without delay through the next statement, but the LHS variables retain their old value. When an end of block is reached, or when execution is delayed the value of T is assigned to the LHS variables.

A non-blocking assignment with `timing_ctrl`: A temporary variable, T, is created and the expression value is stored in T. Execution continues without delay through the next statement, but the LHS variables retain their old value. When the `timing_ctrl` condition or the `repeat_timing_ctrl` condition has been satisfied, the value of T is assigned to the RHS variables.

```
<specify_block>
        -> specify <specify_item>... endspecify
<specify_item>
        -> <specparam_decl>
        -> <path_decl>
        -> <state_dependent_path_decl>
        -> <lvl_sensitive_path_decl>
        -> <edge_sensitive_path_decl>
```

A `specify block` is used to specify timing information for the module in which the `specify block` is used. This information can then be used by either the simulator or by other tools to detect timing violations or to enforce timing constraints. `Specparam` is used to declare delay constants, much like regular parameters inside a module. Paths are used to declare time delays between inputs and outputs.

`State_dependant_path_decls` declares path delays that are conditional on some internal state of the module or UDP. Similarly, level sensitive and edge sensitive path declarations specify delays on paths which relate to specific polarity or transition.

The `specparam_decl` construct contains a list of assignments which are used to declare delay parameters and give them values. In the following we present only a simplified version of the various path declarations.

```
<specparam_decl>
        -> specparam <param_assgn>\','... ';'
<param_assgn>
        -> <identifier> '=' <const_expr> <path_decl>
        -> '(' <inp_list> <connection>
               <outp_list> ')' '='<path_delay> ';'
<connection>
        -> '=>'
        -> '*>'
<path_delay>
        -> <delay_expr>
        -> '(' <delay_expr> ')'
<state_dependent_path_decl>
        -> [if '(' <bool_expr> ')'] <path_decl>
<lvl_sensitive_path_decl>
        -> <state_dependent_path_decl>
<edge_sensitive_path_decl>
        -> [if '(' <bool_expr> ')'] <edge_path_decl>
<path_decl>
        -> '(' <edge_ident> <inp_var> <connection>
               <outp_list> ')' '=' <path_delay> ';'
<edge_ident>
        -> posedge
        -> negedge
```

A `path_decl` is used to specify delays between input ports and output ports. If the connection is '=>' then a parallel connection is established between any bit in the input_list to the corresponding bit in the output_list. If the connection is '*>' then a full connection is established between any bit in the input_list to any bit in the output_list. The `path_delay` can have between one and six constants, but the most common ones are one constant for all delays and two constants for rise delay and fall delay respectively.

A `state_dependent_path_decl` is similar to `path_decl` except that it can be preceded by a conditional clause which controls the delay on the path.

`Level_sensitive_path_decl` and `edge_ sensitive_path_decl` are variations of the `basic_path_decl` which take into account the polarity of the signal and the polarity of the signal transition respectively.

Condensed Language Reference Manual

## Operation

Although the semantics of the language can be explained without referring to the underlying tools that implement it, such as simulators, it is much simpler and more descriptive to assume the existence of such a tool and to describe the semantics in term of this tool. The Verilog simulator that we will assume is an imaginary one, and is not related to any commercial product.

The simulator is an event based simulator. It represents future activities as events that need to be scheduled. The scheduler is the focal point of the simulator. It manages the events, schedules future events and dispatches the execution unit to perform the actions in the events. Some of these actions in turn generate new events which are scheduled to be performed later.

The simulator keeps a global variable which designates the simulation time. This is a 64 bit unsigned integer which is zero at the beginning of the simulation, and which keeps increasing as the simulation progresses. Simulation time is not tied to real time or CPU time. In the rest of the section, time will refer to simulation time unless stated otherwise. Time can be referenced in behavioral calls using the system function time.

Future events are sorted by simulation time. The simulator has the freedom to order the events which are scheduled at the same time in any order. Typical events are: evaluation of a gate or a UDP output, evaluation of the RHS of a continuous assignment and assigning it to the LHS, resuming the execution of a behavioral block, etc. A behavioral block which does not have a time control construct is guaranteed to execute without interruption.

Following are typical operations of the simulator and the event scheduler. When a net on the RHS of a continuous assignment changes, the simulator schedules an event to recalculate the RHS and to assign it to the LHS. When a net that is connected to the input or an inout port of a submodule changes, than the simulator schedules an event to transmit the new value to the inside of the module. When an event variable is triggered, then the simulator schedules all the behavioral blocks that are waiting on the event. When a variable in a block changes, then depending on the polarity of the change, the simulator schedules all the behavioral blocks that are waiting for the change. When a behavioral block is delayed for a fixed period of time, the simulator schedules the resumption of the execution of the block in future time. When a disable statement is executed, the corresponding event is removed from the schedule queue. When an `always` block is finished, it is scheduled to start again at the current time.

At the beginning of the simulation, the simulator schedules all the behavioral blocks, all the gates, and all the UDPs. At each point in time the simulator scans the list of scheduled events for the current time. It performs the events one at a time and removes them from the event queue. When the current time has no more events associated with it, the simulator advances time to the first scheduled event, and the cycle begins again. When no more events exist, simulation stops.

Because the simulator has the freedom to the events within the same time in any order, two different simulators may generate different results. This is an indication that the model was not written well, and that the hardware that it describes has unpredictable behavior. However two runs of the same simulation using the same simulator are guaranteed to give the same results.

# APPENDIX B

# Verilog Formal Syntax Definition

This appendix describes the Verilog HDL syntax. The syntax is given in extended BNF notation. The following conventions are used: items within sharp brackets (<>) are nonterminals. Keywords are given in bold characters. A star (*) indicates that the previous construct can appear zero or more times. A question mark (?) indicates that the previous item is optional, i.e. it can appear zero or one time. One common construct in Verilog is a list of items separated by commas:

```
item, item, .... , item
```

which in BNF can be described as

```
<list_of_items> := <item> <,<item>>*
```

In order to reduce the size of the BNF description, we will omit the definitions of such lists.

```
<verilog_file>
   := <module>*

<module>
   := module <module_name> <module_ports>? ;
      <module_items>*
      endmodule

<module_ports>
   := (<list_of_ports>?)

<module_name>
   := <IDENTIFIER>

<port>
   := <port_expression>?
   |  . <port_name> ( <port_expression>? )

<port_expression>
   := <port_reference>
   |  { <port_reference> <, <port_reference>>* }

<port_reference>
   := <variable_name>
   |  <variable_name> [ <expression> ]
   |  <variable_name> [ <expression> : <expression> ]

<port_name>
   := <IDENTIFIER>

<variable_name>
   := <IDENTIFIER>

<module_item>
   := <parameter_declaration>
   |  <input_declaration>
   |  <output_declaration>
   |  <inout_declaration>
   |  <net_declaration>
   |  <reg_declaration>
   |  <integer_declaration>
   |  <gate_instantiation>
   |  <module_instantiation>
   |  <udp_instantiation>
   |  <always_instantiation>
   |  <initial_instantiation>
   |  <continuous_assignment>
   |  <function>
```

# Verilog Formal Syntax Definition

```
<function>
    := function <range>? <function_name> ;
       <func_declaration>*
       <statement_or_null>
       endfunction

<function_name>
    := <IDENTIFIER>

<func_declaration>
    := <parameter_declaration>
    |  <input_declaration>
    |  <reg_declaration>
    |  <integer_declaration>

<parameter_declaration>
    := parameter <range>? <list_of_assignments> ;

<input_declaration>
    := input <range>? <list_of_variables> ;

<output_declaration>
    := output <range>? <list_of_variables> ;

<inout_declaration>
    := inout <range>? <list_of_variables> ;

<net_declaration>
    := <NETTYPE> <charge_strength>? <expandrange>? <delay>?
       <list_of_variables> ;

    |  <NETTYPE> <drive_strength>? <expandrange>? <delay>?
       <list_of_assignments> ;

<NETTYPE>
    := wire

<expandrange>
    := <range>

<reg_declaration>
    := reg <range>? <list_of_register_variables> ;

<integer_declaration>
    := integer <list_of_integer_variables> ;

<continuous_assignment>
    := assign <drive_strength>? <delay>?
       <list_of-assignments> ;
```

```
<initial_instantiation>
   := initial statement
    |  initial <seq_block>

<always_instantiation>
   := always <statement>
    |  always <seq_block>

<variable_name>
   := <IDENTIFIER>

<register_variables>
   := <IDENTIFIER>

<integer_variable>
   := <IDENTIFIER>

<range>
   := [ <expression> : <expression> ]

<gate_instantiation>
   := <GATETYPE> <drive_strength>? <delay>?
       <gate_instance> <, <gate_instance>>* ;

<GATETYE>
   := and
    |  nand
    |  or
    |  nor
    |  xor
    |  xnor
    |  buf
    |  not

<gate_instance>
   := <gate_instance_name>? ( <terminal> <,terminal>>* )

<gate_instance_name>
   := <IDENTIFIER>

<terminal>
   := <identifier>
    |  <expression>

<module_instantiation>
   := <module_name> <module_instance> <,<module_instance>>*;

<module_name>
   := <IDENTIFIER>
```

```
<module_instance>
    := <module_instance_name> ( <list_of_module_terminals>? )

<module_instance_name>
    := <IDENTIFIER>

<module_terminal>
    := <identifier>
    | <expression>

<named_port_connection>
    := . IDENTIFIER ( <identifier> )
    | . IDENTIFIER ( <expression> )

<statement>
    := <assignment>
    | if ( <expression> ) <statement_or_null>
    | if ( <expression> ) <statement_or_null>
      else <statement_or_null>
    | case ( <expression> ) <case_items>+
      endcase
    | for ( <assignment> ; <expression> ; <assignment> )
          <statement>
    | <seq_block>
    | disable <IDENTIFIER> ;

<assignment>
    := <lvalue> = <expression>

<case_item>
    := <expression> <,<expression>>* : <statement_or_null>
    | default : <statement_or_null>
    | default <statement>

<seq_block>
    := begin
          <statement>*
       end
    | begin : <block_name>
          <block_declaration>*
          <statement>*
       end

<block_name>
    := <IDENTIFIER>
```

```
<block_declaration>
    := <parameter_declaration>
    |  <reg_declaration>
    |  <integer_declaration>

<lvalue>
    := <IDENTIFIER>
    |  <IDENTIFIER> [ <expression> ]
    |  <concatenation>

<expression>
    := <primary>
    |  <UNARY_OPERATOR> <primary>
    |  <expression> <BINARY_OPERATOR> <expression>
    |  <expression> ? <expression> : <expression>

<UNARY_OPERATOR> is one of the following tokens:
    + - ! ~ & ~& | ^| ^ ~^

<BINARY_OPERATOR> is one of the following tokens:
    + - * / % == != === !== && || < <= > >= & | ^ ^~ << >>

<primary>
    := <number>
    |  <identifier> [ <expression> ]
    |  <identifier> [ <expression> : <expression> ]
    |  <concatenation>
    |  <multiple_concatenation>
    |  <function_call>
    |  ( <expression> )

<number>
    := <NUMBER>
    |  <BASE> <NUMBER>
    |  <SIZE> <BASE> <NUMBER>

<NUMBER> is any number made of the following characters:
    0123456789abcdefABCDEF

<BASE> is one of the following tokens:
    'b 'B 'o 'O 'd 'D 'h 'H

<SIZE> is any number of following digits: 0123456789

<concatenation>
    := { <expression> <,<expression>>* }

<multiple_concatenation>
    := { <expression> <,<expression>>* } }
```

```
<function_call>
    := <function_name> ( <expression> <,<expression>>* )

<function_name>
    := <IDENTIFIER>

<delay>
    := # <NUMBER>
    |  # <identifier>
    |  ( <expression> <, <expression>>* )

<IDENTIFIER>
```
   An identifier is any sequence of letters, digits, and the underscore'_' symbol, except that the first character must be a letter or underscore. Upper and lower case letters are considered to be different. Identifiers may be of any size and all characters are significant.

# Index

## Symbols

$display   328
$readmemb   322
$readmemh   322, 324

## Numerics

2-level logic optimization   195

## A

a barrel shift   107
a body   280
a head   280
Adder   109
adder   107
address bus   128
Address sequencer   217
Allocation   240
always   316
AMD2910 microcontroller   217
AMD2910 top-level model   228
Area vs. speed   207
arithmetic operators   204
Associate   148
Associate Cache   148
Associative   148
associative cache   150
associative memory   318
asynchronous systems   265
Automatic test pattern generation (ATPG)   241

## B

Barrel Shifter   110
barrel shifter   332
Behavioral synthesis   193

bidirectional   124, 309
bidirectional port   310
bidirectional ports   179
bus transactions   312

## C

cache   127
cache controller   132
cache driver   148
carry-look-ahead adder   109
case statement   276
checksum   283
Clock Driver   124
Clock Generator   117, 168
clock generator   168
Coding style   191
combinational block   316, 328
compound tasks   330
Condition Codes   112
Consistency between HDL and gate versions   196
Content-Addressable   114
continuous   109
continuously assigned   326
control paths   104
control signals   128
Control Unit   124
Controller   143
controller   266, 269, 274
counter   140
Counter functional model   212
cylinder   278

## D

Data   139
data   281
data bus   128
data communication   164
Data flow diagram   238

371

data port 310
Data RAM 132, 139, 142
data-flow graph 193
Datapath 107
datapath 104
decrement 108
Design Compilation 194
Design debugging 196
design debugging 196
Design optimization 191
design optimization 191
Design partitioning 234
Design partitioning for synthesis 235
Design planning 232
Design structure 236
direct memory acces 271
direct-mapped cache 130
disk drive 278
disk head 278
DMA 272, 277
Don't care for logic minimization 209
drive interface 269
driver module 173
dual UART 164

# E

events 315, 316
exclusive-OR 113
Execution Order 326

# F

fault grading 241
FDC 267
fetch cycle 124
File input/output 236
Finite State Machine (FSM) functional model 213

Finite-state machine (FSM) synthesis 193
floppy disk controller 265
floppy disk drive 265
four-way case 223
four-way multiplexor 223
Functional & timing verification 200
Functional Model of the Single UART 168
Functional verification of RTL design 197

# H

handshake protocol 324
hash collisions 321
hash function 319
hash table access 318
HDL synthesis 193
head 281
high level synthesis 193
hit ratio 127
hold time 114
host interface 269

# I

Incomplete event lists 203
Incrementer 108
Incrementor model 222
index 131
index hole 279
index signal 280, 281
input 309
Interfaces 128
Interleaved Memory 322
interleaved memory 322
interrupt 167, 173

# L

large models    265
Last-in first-out (LIFO)    219
Line Size    150
Logic Optimization    195
Logic synthesis    193
LRU algorithm    318

# M

memory subsystem    318, 322, 330
Microcontroller    217
modeling for synthesis    191
Multilevel logic optimization    195
multiplexor    140
Multiplexor using case    223
Multiplier    112
multiplier    107

# N

nonoverlapping    119

# P

Partitioning    240
Partitioning simulation environment    211
Physical design for synthesis    235
pipeline architecture    312
PLA model    224
Ports    309
ports
    bidirectional    309
    inout    309
    input    309
    output    309
ports, data    310
Post-layout timing analysis    202

post-layout timing analysis    202
Pre-synthesized library    231
Procedural and tabular vectors    199
processor interface    128
program counter    125
programmed I/O    271, 272, 274
Programming language interface (PLI)    241
Project management for synthesis    235

# R

race conditions    277
Random-Access Memory    114
re- use a module    197
Read Hit    142
Read Miss    143
Read Operation    170
read operation    129
read transactions    274
Receive Operation    171
reception    167
register file    116,    173
Register model    219
Register Transfer Level (RTL)    193
Relational operators    204
reset    274
Reset Operation    168
Re-targeting designs    196
re-targeting designs    196
Re-use of stimulus and behavioral models    200
rotate-left    111
RTL assignments    205
RTL synthesis    193

# S

Scheduling   326
sector   280
sector data   283
sector head   283
sector tail   283
sectors   278
setup   114
Setup and Hold Constraints   325
shadow register   310
shift registers   174
shifter   332
shift-left   111
shift-right   111
Simulation of RTL design   241
single UART   164, 168
Single-Phase Clock   118
size for a cache   130
soft macro   231
state machine   140, 317
state table   317
status register   125, 270, 277
structural model   167
Symbol library   232
Synthesis   191
synthesis   191
Synthesis libraries   231
Synthesis partitioning   235
System-level synthesis   192

# T

table lookup   318
Tag   139
tag   131, 140
Tag comparator   132
tag comparator   139
Tag RAM   132, 143
tag RAM   137
tail   280, 281
Technology independent logic equations   194

Technology Mapping   195
Technology translation   195
Test environment   207
test vector generation   331
Test Vectors   329, 331
thrashing   149
timing checker   268, 269
Timing inferrence for synthesis   207
trace file   329
track   278, 280
Traffic Light
    State Machine   215
Traffic light controller   211
transaction   315
transmission   167
Transmit Operation   171
Tri-state   211
Tri-state model   226
Tri-state Out Model   226
Two array indices   238
Two-Phase Clock   118
two-way interleaved memory   322

# U

UART   163
Universal Asynchronous
    Receiver Transmitter   163
user-defined primitive   230
utility tasks   284

# W

wait state   129
Wait State Counter   147
wait states   140
Write   143
Write Buffering   150
Write Hit   142

Write Miss   143
Write Operation   170
write transactions   274
Write-Back   151
write-back   132
write-through   132

# X

X   317,   318

# Z

Z   310
zero delay statement   326

# Order Form

*BookMasters, Inc.*
PO Box 2039, Mansfield, OH 44905
**Phone:** 1-419-281-1802 **Fax:** 1-419-281-6883

```
Name_____Title_____
Company_____
Address_____
_____
_____

Phone_____Fax_____

Purchase Order(if any)_____Dated_____

Charge My __Visa__MC__AmExp:_____Expires_____
```

*Publ. 1:* Digital Design and Synthesis with Verilog HDL (365 pages, hard cover)
*Publ. 2:* Digital Design and Synthesis with Verilog HDL + *Verilog Simulator* under Windows
    (The simulator supports full language, no PLI, limited to 1000 lines of source code)

*Publ. 3:* Quick Reference for Verilog HDL (26 pages, soft cover, spiral ring, ref. by examples)
    Not sold. For a **free** copy call *interHDL*, Inc. at 415-428-4200, Email: info@interHDL.com

**Please send me the following publications:**

|  | Publ. 1 | Publ. 2 | Total |
|---|---|---|---|
| Quantity: | | | |
| Price (use discount table): | | | |
| Sales Tax: (7.75 % in CA) : | | | |
| Total Amount (each publ.): | | | |
| Grand Total (add each col.) | | | US$_____ |

**Discount Schedule Table:** (Price of Publ.+ Shipping & Handling)

| Quantity | **Publ. 1**<br>(Book+S&H) | **Publ. 2**<br>(Book+Simulator+S&H) |
|---|---|---|
| 1 - 9 | ($65+$6)/copy | ($100+$6)/copy |
| 10 - 24 | ($60+$5)/copy | ($95 +$5)/copy |
| 25 - 49 | ($55+$4)/copy | ($90 +$4)/copy |
| 50 - 99 | ($50+$3)/copy | ($85 +$3)/copy |
| 100 plus | ($45+$2)/copy | ($80 +$2)/copy |

**Foreign Buyers:**   Please remit money in **US Dollars via a US bank**.
                     Please add **$25** for **Shipping and Handling**.

**Bookstores/Univ:**  Please call and ask for bookstore **discount**.

**Prepayment Req.:**  All purchases require **prepayment** except for
                     volume purchases by bookstores and companies.